Advance Praise for
The Smartphone Society
by Nicole Aschoff

"Aschoff's analysis of our relationship to our phones is relevant and urgent. She gives us enough context to understand our addictions, our willingness to be surveilled and manipulated, and, better yet, the avenues of resistance against the tech titans that increasingly control our time, attention, and futures."

—CATHY O'NEIL, author of *Weapons of Math Destruction: How Big Data Increases Inequality and Threatens Democracy* and CEO of O'Neil Risk Consulting & Algorithmic Auditing

"An antidote to the typical screen panic, *The Smartphone Society* reframes our phones as a new frontier of American life. It's a useful read for anyone worried about how we live with technology, and that should be all of us."

—MALCOLM HARRIS, author of *Kids These Days: The Making of Millennials*

"In *The Smartphone Society*, Nicole Aschoff gives us fresh insight into how the device and our everyday lives have morphed into one another. She considers the good and the bad, and helps us to understand how the smartphone has reshaped society in innumerable ways. With accessible prose, she looks into selfies and social media, politics and protest, profit and women's unpaid work. It is a cogent read in the era of the smartphone."

—RICH LING, Shaw Foundation Professor of Media Technology, Nanyang Technological University

"*The Smartphone Society* is not your average tech book—it's one about who controls our future. In our New Gilded Age, the book persuasively argues, it'll be either be dictatorial tech giants or the democratic power of free citizens. I know which outcome I prefer, and I can think of few better intellectual defenders of a better, more just future than Aschoff."

—BHASKAR SUNKARA, editor and publisher, *Jacobin* magazine

"*The Smartphone Society* pierces the fog of the Silicon Valley fantasy, showing us how these little computers in all our pockets control our lives for profit—but also how they open new paths to justice. Nicole Aschoff has given us that rare book, packed with insights and written with verve. I will never look at my smartphone the same way—and after reading *The Smartphone Society*, neither will you."

—JASON W. MOORE, professor of sociology
and author of *Capitalism in the Web of Life*

The Smartphone Society

Also by Nicole Aschoff

The New Prophets of Capital

The Smartphone Society

Technology, Power, and Resistance in the New Gilded Age

Nicole Aschoff

Beacon Press, Boston

BEACON PRESS
Boston, Massachusetts
www.beacon.org

Beacon Press books
are published under the auspices of
the Unitarian Universalist Association of Congregations.

23 22 21 20   8 7 6 5 4 3 2 1

This book is printed on acid-free paper that meets the uncoated paper
ANSI/NISO specifications for permanence as revised in 1992.

Text design and composition by Kim Arney

Library of Congress Cataloging-in-Publication Data

Names: Aschoff, Nicole Marie, author.
Title: The smartphone society : technology, power, and resistance in the
  new gilded age / Nicole Aschoff.
Description: Boston : Beacon Press, 2020. | Includes bibliographical
  references and index.
Identifiers: LCCN 2019026269 (print) | LCCN 2019026270 (ebook) |
  ISBN 9780807061688 (hardcover) | ISBN 9780807061961 (ebook)
Subjects: LCSH: Smartphones—Social aspects—United States. | Online social
  networks—United States. | Cell phone services industry—United States. |
  Mobile computing—United States.
Classification: LCC HE9713 .A83 2020 (print) | LCC HE9713 (ebook) |
  DDC 303.48/33—dc23
LC record available at https://lccn.loc.gov/2019026269
LC ebook record available at https://lccn.loc.gov/2019026270

Dedicated with love and affection to
Parmender P. Mehta
(1953–2019)

Contents

Introduction

"I've lost my phone!" a woman wailed. Choking down panic, she realized she'd left it on the backseat of the taxi on the way to the airport. Bystanders quickly mustered to help, calling the lost phone, offering advice and moral support. The woman was Lucy Kellaway, a *Financial Times* reporter who later recounted her ordeal in a fluffy piece for the newspaper. Kellaway observed that "on the scale of human calamities, losing your phone is now seen as up there with cardiac arrest," and that even with her wallet and laptop still in hand, as she sat in her hotel room that evening, she felt "all wrong: exposed and vulnerable" because her smartphone was hundreds of miles away.[1]

Most people have lost their phone at some point, or have feared they had, and have experienced that moment of sweaty, heart-pounding dread—but Kellaway's story is illuminating all the same. It neatly captures our surprise and unease at how attached we've become to pocket computers. We simply can't imagine life without them.

Whether we're fine with this attachment, resolvedly not thinking about it, or trying to delink with a new "dumb phone," the reality is clear: in just ten years, smartphones have become woven into the fabric of everyday life. Smartphones have catalyzed rapid shifts in how people communicate, fall in love, raise their families, and burnish their social status. They have facilitated the creation of new jobs and new ways of working; they have given rise to new consumption patterns; and they are at the center of new visions about democracy, politics, and the future.

Yet despite momentous technological advances, the present moment is fraught with contradictions. Observers liken it to the late-nineteenth-century Gilded Age, a time of robber barons, staggering inequality, and seismic technological and social shifts. In this New Gilded Age, life seems more precarious than ever. Society is riven by racism, sexism, and xenophobia. Regular folks feel lucky to get and hold on to a job in an economy dominated by superstar firms. The powerful unions and social safety net that once protected workers and their families are a distant memory. As the gap between rich and poor yawns ever wider, anxiety about a "robot future" in which technology replaces the livelihoods of ordinary people abounds.

The changes we've witnessed since Steve Jobs ever so casually pulled an iPhone out of his pocket at Apple's 2007 Macworld meeting in San Francisco feel monumental and unprecedented—and they are.[2] Yet a glance to the past shows that we've been here before. In their landmark study *Middletown: A Study in Contemporary American Culture*, the sociologists Robert and Helen Lynd described a similar moment of flux: 1920s America and its emergent love affair with the automobile.[3]

By the end of the roaring twenties, the automobile had become central to the lives of "Middletown" (actually, Muncie, Indiana) residents, an "accepted essential of normal living" and an "important criterion of social fitness."[4] Homes built during the period no longer contained formal parlors because unmarried daughters now socialized with their beaus on unchaperoned "dates" out in the car—a source of both fierce disagreement between parents and children and growing moral panic over "sex crimes" committed in automobiles.[5] Young men whose families didn't own a car were excluded from social clubs, and families were rumored to purchase cars to help their children fit in. Walking for pleasure or bicycling marked one for abuse from young boys known to shout, "Aw, why don't you buy a machine!"[6] Leisurely weekend lunches followed by afternoons socializing on the front porch with extended family and neighbors had given way to Middletown housewives preparing an informal "bite" so Mom, Dad, and the kids could "get out in the car."[7] Sunday car treks also encouraged skipping church—another threat to "group sanctioned values."[8]

Middletown pastors weren't the only ones complaining of a loss of devotion. Shopkeepers groused that people spent every spare dime on their automobiles. Working families devoted a week's pay every month to financing the family vehicle. And when work dried up, as it regularly did, residents declared they'd go without new clothes and even food rather than "give up the car."[9] Cars also helped Middletowners cope with painful bouts of unemployment. For the first time, men could commute to other towns to find work and they were doing just that, creating a "migratory laboring population."[10] This severing of the need for proximity between home and work increased home ownership, road construction, and car-related industries, but gravely worried city boosters and business owners who saw a migratory population as less moral, less politically engaged, and less likely to spend money where they lived.

But working families had few options. The assembly lines that men toiled on to build the cars and the parts that went into them changed the way work was organized and, by extension, how ordinary people made a living. Eighteen-year-old "boys" were suddenly earning a "man's wages" (which, alongside the entry of women into the paid workforce, encouraged early marriage), while men in their forties and fifties were routinely let go, deemed too old for the assembly line.[11]

Nineteen twenties America may be a bygone era, but the parallels between the moment described by the Lynds and the present are striking. The automobile, like the smartphone today, brought both a sense of wonder and intense anxiety. Like those Middletowners, we are in the midst of society-wide social, economic, and political change, and a new machine is once again at the center of it all: the smartphone. If the automobile was the defining commodity of the twentieth century, the smartphone is the defining commodity of the twenty-first.

A New Tether

More than 80 percent of American adults own a smartphone. The average iPhone user checks her phone eighty times a day—that's about thirty thousand times between one birthday and the next.[12] In the United States, where the modern smartphone was born, roughly six out of ten adult smartphone users sleep next to their phones.[13] If we wake up in the middle of the night we might quickly scroll through our messages. Many of us don't eat breakfast or go for walks without our phones—not even a walk to the bathroom.

Americans are not alone in this behavior. In Germany the word "smombie," a portmanteau of "smartphone" and "zombie," was coined to describe the prevalence of zoned-out smartphone users shuffling around town. Smombies are traffic hazards, prompting Augsburg, a city northeast of Munich, to embed warning lights in sidewalks to alert smombies of approaching streetcars. Chongqing, China, has toyed with designated walking lanes for phone users, modeled on an earlier experiment in Washington, DC.[14]

Whether we're crammed on the subway, watching our kid's Tae Kwon Do practice, or fidgeting at a stoplight, most of us are on our smartphones, which have rapidly overtaken older, limited-feature cell phones.[15] Our umbilical attachment to our smartphones means they're always at the ready to film happenings, whether it's a clumsy lion falling into a pond in a German zoo or the local police brutalizing a member of the community. Look at the crowd at a Taylor Swift concert—nearly everyone is filming the show. Ponder the photos of the worldwide vigils that followed the 2016 Pulse nightclub shooting in Orlando, Florida—phones are aloft, flashlights on, in a show of respect and solidarity.

We fill in the gaps and spare moments of our days socializing with each other through our smartphones. We send trillions of texts (the vast majority going to family and close friends), swipe through Tinder profiles searching for our soul mate, and video-chat with cousins in India. Indeed, Indians use so much bandwidth sending "Good Morning!" messages and photos on WhatsApp that their phones routinely run out of space and freeze.[16]

Smartphones are quickly becoming our "everything" device.[17] Nowhere is this clearer than in China, where the vast majority of the country's eight hundred million internet users connect to the web through their *shou jis*, or "hand machines."[18] The mammoth e-commerce conglomerate Alibaba allows users to log in to their accounts with a selfie and pay for nearly everything with Alipay, Alibaba's mobile payment system; even homeless people use Alipay to accept donations. Users of rival conglomerate Tencent's WeChat app—a multipurpose app for social media, messaging, and mobile payments—send each other digital "red envelopes" (the traditional way to send a gift in China), hail a cab, book doctors' appointments, or buy noodles at tiny street stalls, all using the same app. Urban Chinese can go weeks without pulling out their wallets; indeed, paying in cash risks marking oneself as *tu*—Mandarin for provincial or unsophisticated.

Given the nearly universal use and acceptance of smartphones, commenting on how social norms have been reorganized around, or created by, smartphones might seem a bit like telling an "I lost my phone" story—relatable but not remarkable. Our expectations about what we should get from, and how we interact with, digital technology have shifted so quickly and dramatically since 2011—the year the Pew Research Center conducted its first stand-alone study of smartphone ownership in America and found that roughly a third of American adults owned a smartphone—that we often breeze over how norms and behaviors have changed.[19] As the nineteenth-century historian Charles Francis Adams Jr. said of the newly omnipresent railroads, "Whatever constantly enters into the daily life soon becomes an unnoticed part of it."[20]

On the most basic level, we expect to have the internet and an extremely accurate GPS (Global Positioning System) at our fingertips at all times. Suddenly starving on a road trip, we might google the nearest drive-thru. Heading to a museum on the weekend or the new guy at work's barbecue, we don't even look up the directions until we're walking out the door. Remembering that our credit card payment is due while we trudge on the treadmill at the gym, we pay it immediately with a few sweaty taps.

We also expect to be able to reach others directly, or at least send them a message that they will receive, anytime, anywhere. Whether they are across town, in another state, or in a different country, we increasingly expect to be able to see them, in real time, on video. Although advancing text and chat capabilities have also made using mobile phones more unobtrusive—it's relatively rare to encounter the loud phone talker, a public menace in the late nineties and early 2000s—smartphones are constantly visible, ready to be picked up.

While at work, people are texting the babysitter and getting back a snap of the munchkin on the swing; they're WhatsApping their girls-night-out group and Bumbling a divorced accountant in Hoboken; they're drooling over a '67 Airstream on Craigslist and using Petzi, a smartphone-controlled pet cam, to check up on their pup. And doing all this while quietly plugging away in their cubicle. Conversely, we scarcely bat an eye at checking work emails or texting with our boss while waiting in line to buy movie tickets or making a Costco run. We've come to accept that teens' worlds are one part analog (connoting a material connection to a person), one part digital. Social networks stretch from school to the bedroom and back again, while mobile gaming creates dynamic, one-to-one relationships between young people on opposite sides of the world.

Turning over a new smartphone feels like touching the future. We now carry with us, everywhere we go, a powerful minicomputer that connects us to each other, and to cyberspace, perpetually and powerfully. Nonetheless, it is worth remembering that our smartphones are in many respects a mash-up of existing technologies and behaviors packed together in a novel configuration.

Time magazine named the PC "person of the year" in 1982. Hardware, touch screens, radios, processors, antennas—much of the technology essential to our smartphones predates the advent of the actual smartphone. Steve Jobs's team at Apple figured out how to refine and combine these technologies into one device, creating the iPhone.[21] Many software concepts also predate the smartphone. For example, mobile payment systems spread widely in poor countries through cell phones. Kenyans developed the popular M-pesa mobile money transfer system, which lets users lacking bank accounts send cash to friends and family; in early versions senders deposited cash at an agent's shop, then used a text message to send the funds.

American adults may remember the buzz around former president Barack Obama and his Blackberry. Obama used his Blackberry to keep up with emails and refused to part with it upon taking office, a new development for a security-obsessed White House. Email created the ethos central to social media, providing new ways of interacting and sustaining relationships, particularly with people who lived far away. Phone calls were often expensive and emotionally consuming; with email you could effortlessly dash off a line or two, updating friends and family on life's major and minor events.

Social media also predates smartphones, but today they have become inseparable from smartphones, along with their norms and behaviors. Indeed, many of the norms and expectations we associate with the smartphone were already forming with plain old cell phones. Texting took the world by storm in the late 1990s. Worldwide, people sent a trillion text messages in 2005. As French sociologist Christian Licoppe notes, SMS—short message service—allowed users to "decontextualize their interactions," encouraging spontaneity and impulsivity.[22] Cell phones created what Licoppe calls "connected presence"; for the first time, we could call people directly, not just the location we hoped they'd be in.[23]

"Wherever/whenever communication," as communications scholars Rich Ling and Jonathan Donner call it, represented "a breach in the way that we generally coordinate[d] interaction." The "finer-grained form of synchronization" facilitated "more flexible forms of social coordination," raising new questions about social cohesion and the tensions between in-person communication and "mediated" interactions that remain unresolved today.[24]

Yet, smartphones are much more than an accumulation of improvements in hardware and software into a pocket-sized device that we spend too much time looking at. They represent something entirely new. When we pick up our phones, our taps and swipes engage not only a system of hardware and software, but also something much bigger—a set of institutions, relationships, and networks that have come to define modern society. Smartphones are catalyzing new ways of being and interacting, new ideas about identity and the nation—a qualitative shift from ten years ago. Our pocket computers are at the center of a profound reconfiguration of familiar values, of the boundary between the public and the private. In the past decade, our smartphones have become a new tether, tying us to each other, to the digital, and to our broader socio-economic system in unprecedented ways. We've become a smartphone society.

Smartphone Skeptics

While we're rapidly fleshing out the contours of a new smartphone life, many observers are deeply uneasy about the society we're creating for ourselves. Sherry Turkle, a science and technology professor at MIT and a leading critic of our smartphone society, believes a new "silent spring" (an allusion to Rachel Carson's term for the silencing of birdsong due to DDT)—is on the horizon. Only this time it isn't DDT that is killing all that is beautiful; it's our smartphones. According to Turkle, we've become so accustomed to being connected that we're terrified of being alone—any spare second and we're on our phones, silently swiping and tapping away in phone world. The upshot, says Turkle: we are "less empathetic, less connected, less creative and fulfilled. We are diminished, in retreat."[25]

Eric Pickersgill, a North Carolina–based artist, expresses Turkle's critique visually in his photo series "Removed." Pickersgill's photos show people mesmerized by their tiny screens, or at least by the space where their device would be. The artist poses his subjects with their phones but removes the devices just before making the exposures. Pickersgill says, "We have learned to read the expression of the body while someone is consuming a device and when those signifiers are activated it is as if the device can be seen taking physical form without the object being present."[26] The viewer is left with eerie images of lovers, families, friends, staring down at their empty hands, oblivious to the ones closest to them, in effect alone.

Turkle's and Pickersgill's critiques are part of a sprawling range of critiques of smartphones that tap into a wide range of concerns. Several studies find a link between smartphones and cancer.[27] Nicholas Carr, a well-known writer on technology, is concerned that our hand machines are making us dumb. Smartphones are "an attention magnet unlike any our minds have had to grapple with before," Carr says, and our brains don't seem to be up to the task.[28] Natasha Dow Schüll, a cultural anthropologist, says smartphones—operating like little pocket slot machines, inviting us to swipe, tap, swipe, again and again, rewiring our neurological reward centers in the process—put us into the "machine zone."[29]

Fantastic Beasts star Eddie Redmayne recently joined the ranks of smartphone refuseniks, switching to a "dumb phone" that could only be used for voice calls and texting. Redmayne said the decision was "a reaction against being glued permanently to my iPhone during waking hours."[30] The Austrian designer Klemens Schillinger created the Substitute Phone, a smooth wooden rectangle the same shape and weight as a smartphone, to help smartphone addicts give up their phones. In the space where your thumb would normally be mindlessly swiping through your Instagram feed is a line of smooth wooden roller balls.[31]

Other critiques focus on how smartphones are damaging social cohesion and our notions of truth and trust. Sociologist Chris Rojek says our mediated exchanges—FaceTiming with friends, for example—"are not natural."[32] Bernard Harcourt, a critical theorist at Columbia Law School, is one of many skeptics who say our smartphone addiction has left us vulnerable to giant corporations such as Google and Facebook who capitalize on and reinforce our attachment to our hand machines to steal our data, spy on our every blink and whisper, and manipulate our beliefs and ideas with fake news and rigged news feeds.[33] These developments, argues Shoshana Zuboff, professor emerita at Harvard Business School, are part of a new mutant form of capitalism called "surveillance capitalism."[34]

Worse, critics say we've allowed our phones to victimize our children. Catherine Price, in a piece for the *New York Times*, confessed, "I recently had a baby and was feeding her in a darkened room. It was an intimate, tender moment—except for one detail. She was gazing at me . . . and I was on eBay, scrolling through listings for Victorian-era doorknobs."[35] Cyberbullying, predators, porn: Child psychologist Richard Freed's take is similar to the message of a famous antidrug PSA from the 1980s showing Dad cracking an egg into a sizzling frying pan, his gruff voice declaring, "This is your brain on drugs." Freed says that as far as kids and smartphones go, moderation sounds good, but it doesn't apply. He likens parentally limited smartphone use to giving kids a little bit of drugs or alcohol.[36] Psychologist Jean Twenge says smartphones have destroyed a generation of young people: "It's not an exaggeration to describe iGen as being on the brink of the worst mental-health crisis in decades. Much of this deterioration can be traced to their phones."[37]

On the Other Hand

Scary critiques of smartphones increasingly seem to permeate popular narratives. Everywhere we turn online, features with impressive graphics and earnest podcasters mull over the dangers of our hand machines. Are we crazy to be carrying these things around in our pockets? Or are we in the midst of a moral panic in which we project broader fears about social change onto an easily identifiable target?

These are difficult questions to answer, not least because there is so much disagreement about the impact of smartphones on individuals and society more broadly. Anthropologist Judy Wacjman disagrees with Turkle, arguing that "smartphones should be regarded as another node in the flows of affect that create and bind intimacy."[38] Ling, in a pathbreaking work on mobile telephony, argues that the mobile phone "extends the reach of parents, children, and friends" and actually "seems to result in stronger internal group bonds."[39] Scholars studying social cohesion in Korea, France, and Japan reported similar findings. Clive Thompson, in *Smarter Than You Think*, argues that digital technology is making us smarter, not dumber.[40] Daniel T. Willingham, a cognitive scientist at the University of Virginia, says that while smartphones are certainly distracting they are not "eating away at our brains."[41]

What about the children? The American Academy of Pediatrics recently revised its recommendations, moving from a total ban on screen time for children under two to saying a moderate amount of high-quality media can be beneficial for young children.[42] Psychologist Richard A. Friedman questions the assertion that American teenagers are suffering from an anxiety epidemic caused by digital technology, pointing instead to a "cultural shift toward pathologizing everyday levels of distress."[43] Andrew Przybylski, a scholar at the Oxford Internet Institute, and Netta Weinstein, a psychologist at Cardiff University, conducted a study of British youth and smartphones, using a representative sample of more than 120,000 UK adolescents, and found that moderate amounts of screen time "are unlikely to present a material risk to mental well-being" and "may be advantageous in a connected world."[44]

Moreover, smartphones bring many good things. While many of us don't stretch the capacities of our phone beyond posting snarky Twitter commentary or exchanging a zillion texts with someone whom we have only a mild desire to meet for coffee, some have pushed smartphone use much further. Jean Judes, executive director of Beit Issie Shapiro, an Israeli nonprofit that advocates for people with disabilities, points out that features central to the smartphone experience—the vibrate feature, predictive text, and touch screens—were originally developed for people with disabilities.[45] BIS has developed a new app called Open Sesame that lets users with spinal cord injuries, multiple sclerosis, neuromuscular diseases, amputations, and other health challenges operate their mobile phones without needing to touch them. The phone's front-facing camera tracks their head commands—slight movements signal a tap or a swipe—allowing users to search the web, text, use social media, and more.[46]

The BIS initiative is just the tip of the iceberg. A man in England with diabetes created smartphone attachments to read his blood sugar. Physicians use apps that display customized body images, making it easier for them to talk to patients about their physical condition. And in poor countries with few doctors, smartphones enable midwives and rural health professionals to consult with experts hundreds of miles away. For example, in Nigeria doctors are using smartphones to tackle tuberculosis. The phones streamline and improve the collection of TB data, making it faster and easier for policymakers and health managers to improve care.

Google Translate can translate "Where is the bathroom?" into thirty-two pairs of utterances and into displayed text in over a hundred languages. Hundreds of millions of people in sub-Saharan Africa rely on mobile money accounts, which allow people lacking bank accounts to easily and securely send money to relatives far away. Business-to-business apps help small businesses survive, while farmers rely on smartphones to give them accurate crop prices, reducing abuse by dishonest middlemen.

More broadly, smartphones connect billions of people to the internet. Those of us with high-speed home connections and personal computers take this access for granted, but roughly 60 percent of the world's people, including roughly one in five Americans, rely on smartphones as their only access point to the web.[47] In this way smartphones make accessible a world of knowledge and connection that would otherwise be inaccessible. This is incredibly important, particularly for individuals who feel isolated in their communities by poverty or difference. For example, for LGBTQ teens living in conservative families, communities, or countries, smartphones provide crucial access to a safe space for experimentation and acceptance.

Amid the storm of criticism of smartphones and digital tech more broadly it is also helpful to put our fears about smartphones into historical context. People have always been anxious about new technology. Sociology as a discipline was founded during the late nineteenth century by scholars— such as Karl Marx, W. E. B. Du Bois, Max Weber, Jane Addams, and Emile Durkheim—who were puzzling out how society (a new concept in its own right) was adapting to industrialization, machines, and technology. Their predictions were often dire.

New inventions such as the telephone and telegraph "annihilated space with time," Wacjman notes, but left many wondering whether "science and technology were advancing faster than the ability of human society to cope."[48] Historian Lynn Spigel reminds us that popular writers of the time (Mark Twain, Edward Bellamy, Henry David Thoreau) feared "that people would become prisoners to machines, sacrifice romance for scientific utopias, or trade the beauty of nature for the poisonous fruits of industrialization."[49]

Americans agonized over what machines were doing to their minds. Writing on the first Gilded Age, historian Jackson Lears describes a "powerful feeling among the middle and upper classes—a sense that they had somehow lost contact with the palpitating actuality of 'real life.'"[50] Philosopher Walter Benjamin, like Turkle today, lamented how in the hustle and bustle of modern life, we no longer let ourselves be bored or slip into a deep state of relaxation necessary for creativity: "Boredom is the dream bird that hatches the egg of experience. A rustling in the leaves drives him away. His nesting places . . . are already extinct in cities and are declining in the country as well. With this the gift for listening is lost and the community of listeners disappears."[51]

As the twentieth century dawned, our uneasiness about technology persisted. New innovations such as the automobile, with its possibilities for facilitating sexual encounters and a footloose life style, and the radio, with its hold over its purportedly helpless captive audience, kept us up at night. Then came television, a technology that seemed designed to spark both wonder and panic.

Media theorist and cultural critic Neil Postman has argued that with television came the end of childhood: "Not since the Middle Ages have children known so much about adult life as now."[52] Fredric Wertham, a psychologist who declared in the 1950s that comic books were destroying a generation, also warned parents that TV characters such as "Captain Video and Superman would . . . turn children into violent, sexually 'perverse' adults."[53] Sociologist Robert Putnam located television at the center of a disintegrating society. Instead of chatting with the neighbors, going bowling, or participating in civic activities, Americans were sitting at home glued to the boob tube.[54]

Critiques of technology have been a persistent element of public discussion for at least a hundred and fifty years. Given this historical context, are our fears about our emergent smartphone society overblown? Absolutely not. But they are often muddled and misdirected. To think productively and clearly about the central issues plaguing our smartphone society—toxic social media, monopolistic tech companies, social divides, government surveillance, decaying democracy, alienation—we first need to properly orient our critiques, questions, and analysis.

Framing Matters

A big stumbling block is the way analyses of smartphones are often framed. Many studies, whether they end up presenting a positive or a negative view of smartphone technology and its relationship to society, start with the assumption that the phone itself is the primary change agent. Often, the question posed is whether the smartphone is the cause of a particular outcome or phenomenon.

This is understandable, given the sudden ubiquity of smartphones. But, as intuitive as it seems to put smartphones in the driver's seat of social change, it is incorrect; it reproduces the common error of technological determinism—the belief that technology orders society, not the other way around. Arguments about whether smartphones cause or don't cause this or that outcome situate technology outside social relations, positioning it as an isolated and independent force acting on preexisting social relations. This is not the case; technology and social relations evolve together. Indeed, technology embodies social relations—how smartphones are developed and used reflects society, its prevailing ideas, cultural norms, economic imperatives, and divides.[55]

Contemporary American society is a fraught landscape. The richest 1 percent of the population earns eighty-one times the bottom half and owns 40 percent of the country's wealth.[56] MIT economist Peter Temin describes a "dual economy" in the United States: one track for subsistence workers and another for upwardly mobile, skilled workers and elites.[57] National data supports Temin's characterization. The bottom seven deciles have seen annual hourly wage growth of 0.5 percent or less since 2000, and 40 percent of American adults would be unable to cover an unexpected $400 expense. In 2018, household mortgage, student, auto, and credit card debt increased for the fifth year in a row, yet the majority of Americans don't have enough money in the bank to cover one month of household expenses.[58] Despite low official unemployment, one in five children live in families eking out an existence on an income below $25,750, which is the federal poverty line for a family of four.[59]

These economic divides drive a broad crisis of faith in the status quo—a legitimacy crisis that germinated in the aftermath of the 2008–10 financial crisis and its resolution skewed in favor of elites. Increasingly, American society is plagued by a widespread sense of alienation and despair evident in myriad ways: the nationwide opioid epidemic, rising rates of suicide, and historically low levels of trust in the US government. According to a study by the Pew Charitable Trust, only 18 percent of Americans say they trust the government to do what is right "just about always" (3 percent) or "most of the time" (15 percent).[60] The Rand Corporation's recent report *Truth Decay* found that popular distrust and wariness had spread far beyond people's feelings about the government. The report's authors described "declining trust in formerly respected sources of factual information" and growing "uncertainty and anxiety" throughout the United States.[61] In other spheres, belief in education as an escape route out of poverty is dampened by the resegregation of public schooling and skyrocketing college tuition, while fresh reports of widespread future job loss due to technology and artificial intelligence seem to appear on a daily basis.

Deep economic divides and a growing sense of alienation and despair among ordinary citizens have bred political polarization and an upsurge of social unrest in the decade following the financial crisis. Organizing around issues of police violence, gun law reform, women's rights, climate change, a living wage, and healthcare reform has picked up steam as new social movements have emerged. In 2016 we witnessed an unprecedented display of shifting political sentiments, culminating in the election of Donald Trump—a candidate far beyond the pale, relative to the political consensus that has reigned since the late 1980s.[62]

In 2010 only 3 percent of American adults owned a smartphone; by 2013 it was more than half; and by 2019 smartphone ownership had climbed to more than 80 percent. When examining the central place that smartphones have come to occupy in this short span of time, it is imperative that we situate our hand machines within the contours of a dynamic—and in the present moment, fraught—political, economic, and social landscape. Our smartphones are not acting upon, or molding, this landscape from the outside. They are part of it. Our pocket computers reflect the shape of society.

By the same token, however, it is facile to deny the significance of smartphones in and of themselves, to dismiss their transformative qualities as nothing but an artifact of deeper social structures. As Langdon Winner, a renowned scholar of science and technology, observed in 1980, "This conclusion offers comfort to social scientists: it validates what they had always suspected, namely, that there is nothing distinctive about the study of technology in the first place." If there's nothing distinctive about technology, there is little need for new analysis. The standard models of social power—"interest group politics, bureaucratic politics, Marxist models of class struggle"—suffice.[63]

But of course there is much that is distinctive about our pocket computers. Smartphones are at the center of modern daily life, and their significance stems in large part from their unique qualities. To make sense of the present moment we need new models of social power and fresh perspectives. We need, as scholar-activist Harry Braverman advocated, to pair "concrete and historically specific analysis of technology and machinery on the one side and social relations on the other," and to examine "the manner in which these two come together in existing societies."[64]

The Smartphone Society takes a page from Braverman, developing new models and a fresh perspective, showing readers how our phones are not the driver of change in contemporary society but rather, are a window and a tool. The ways people, companies, and governments produce, design, regulate, use, dispose of, and think about smartphones offer a lens into the momentous shifts we're witnessing. They help us decipher the emergence of a new digital-analog reality in which daily life has become seamlessly interwoven with digital and analog threads. In this book I peer through this lens to examine this new reality.

A Few Caveats

This is not a history book. I do not detail the birth of the smartphone, or the internet; I do not recount the rise of the tech giants that dominate modern society. Nor is this a business book, extolling the wonders of platform business models or killer apps. *The Smartphone Society* is a book about society—in particular, American society and the transformations it is grappling with since the birth of the iPhone in 2007.

Smartphones are used worldwide, and many countries are witnessing the emergence of their own smartphone societies. Our pocket computers are not used in uniform ways around the world, however. The same Android handset embodies different cultural meanings, political relationships, and economic arrangements depending on whether it's used in Japan or Jakarta. These nuances and differences are far too great to capture in one book. That said, it will quickly become apparent to readers that the dynamics shaping America's smartphone society are global; and, on the flipside, that the practices and ideas guiding smartphone use in the United States, whether in the sphere of business, politics, or love, have a deep impact on people in other countries.

Thus, this book has an encompassing eye. I am deeply invested in how our personal troubles, choices, and questions are related to, and indeed inseparable from, much bigger issues and structures. *The Smartphone Society* examines how our individual engagement with our hand machines, for better and for worse, is part of a societal transformation—a new era of tension, uncertainty, and possibility.

One final note: This is a book for everyone. I present an analysis of the pressing issues of our time—mediated interaction, oppression, surveillance, algorithmic management, neoliberal ideology, financialization—in a language and framing accessible to all readers. I adopted this approach consciously, guided by the conviction that the future of our smartphone society should be widely discussed and democratically decided. The challenges of this moment are too important to be left to software engineers, academics, and venture capitalists.

Chapter 1

New Divides
(or, Old Divides Made New)

Tanya Marshall lives in Cambridge, Massachusetts. For Tanya, like many of us, days and evenings are packed: she's a parent, a teacher, and an active member of her community.[1] Like most parents, Tanya worries about her children when she's not with them. She uses her phone to keep close tabs on them throughout the day, particularly her son James, who is just finishing high school. James texts her when he's leaving home, school, or extracurricular activities, and tells her which way he's walking so she'll know how long it will take him to get where he's going; if she's at work, he'll text her to let her know he has arrived home.

But unlike many parents, Tanya has an added worry: the police. Tanya is Black and worries that a police officer might harm one of her children. Tanya's fears were confirmed a few years back when her son was a high school freshman. James and a few classmates were on a Red Line train on the "T," the Massachusetts subway system, headed home after football practice, when six police pulled them off the train at Harvard Station. The officers made the boys sit on the platform and began shouting at them, accusing them of breaking a window on the train. James and his friends protested that it wasn't them, but the police continued aggressively interrogating and berating them. The boys, just fourteen at the time, were terrified.

Tanya has a house rule: If her children have an interaction with a police officer they are to call her immediately, while it's happening, so that she can intervene. She knows other Black parents who have the same rule. As James sat on the platform, he and a friend surreptitiously dialed their moms, keeping their phones in their pockets because the policemen had already taken the other boys' phones away. But James was too afraid to take his phone out of his pocket, so Tanya could only listen in agony, praying the officers wouldn't do anything more severe than yell.

The ubiquity of the smartphone encourages an impression of equality, of sameness. Step into a coffee shop or doctor's waiting room and look around—everyone is peering down at their tiny screens. Elites and ordinary folks, teenagers and senior citizens, nearly all have a phone, and most are doing similar things—chatting with friends and family, entertaining themselves with gossip, games, and news. Often we're doing *exactly* the same things on our phones: 1.5 billion of Facebook's users are daily visitors to the site; Luis Ponsi's *Despacito* video got more than 5.5 billion views in 2018.

But as Tanya's experience demonstrates, in the United States these similar behaviors and cultural preferences are transposed onto a deeply divided society. We're divided by race and ethnicity, by whether we live in the country or the city, by gender and sexual preference, by religion, and by income and wealth. Smartphones both reflect these deep divides and reconfigure them in novel ways.

A Tool for Justice

James was deeply shaken by his run-in with the police, breaking down in tears once he reached home. The situation could have turned deadly in an instant. In Oakland, Oscar Grant III was pinned down on a Bay Area Rapid Transit train platform and shot in the back by a BART police officer after a fight had broken out on the train.[2] In Chicago, seventeen-year-old Laquan McDonald was shot in the back walking away from a policeman.[3] Police shot and killed fifteen-year-old Jordan Edwards as he and his older brother tried to drive away from a house party that had gotten out of hand in Balch Springs, Texas.[4] Tamir Rice was twelve when Cleveland cops rolled up and shot him through the window of their police cruiser, in broad daylight in a public park.[5]

James's encounter captures the day-to-day reality for many Black people in America and fuels a pervasive sense of anxiety and caution, instead of a feeling of security and trust, toward the people sworn to protect our communities. When James finally arrived home, Tanya cried, too, terrified by what could have happened. Once again, a parent and child were forced to have "the talk" about how police interactions in America can quickly spiral into violence. Tanya reemphasized to her son that he needed to have his phone within reach at all times. Even though she wasn't able to intervene that day, Tanya sees smartphones as an important tool to keep her children safe from police.

Tanya's concerns are not new. Black and brown Americans have long feared that interactions with police can quickly turn deadly, but smartphones have come to play a central role in this longstanding antagonism. Increasingly, people of color in America use their phone as a tool to protect themselves and others in their communities—to film interactions with police—in the hopes of increasing their safety or, at the very least, their chances of getting justice after the fact.

"The Counted," a project of the *Guardian* newspaper, gives a sense of the disproportionate violence visited upon Black Americans by the police: In 2015, 7.79 Blacks per million Black people were killed by police; in 2016, the figure was 6.66 per million. The figures for whites were 2.95 (2015) and 2.9 (2016) per million whites.[6] In absolute terms many more white people than people of color are killed by law enforcement officers every year; poor whites and those suffering from mental illness are disproportionately likely to die in encounters with police. Yet, relative to Black and Indigenous Americans, white people are much less likely to be killed or incarcerated.[7] People are using their smartphones to document this dynamic.

Before smartphones, national discussions about police behaving violently often circled back to Rodney King. King was beaten nearly to death by Los Angeles police officers in 1991. The beating wasn't a rare occurrence, but it is burned into collective memory because a stranger filmed the assault on King with a camcorder and gave the tape to the local news station. Until recently, however, the filming of violent interactions with cops remained a rarity. Police officers shot forty-one bullets into Amadou Diallo in New York City in 1999, but no one recorded his murder.

Smartphones have made this violence visible. When people think about police violence today, they might think of the case of Walter Scott. In April 2015, Feidin Santana pulled out his phone to film an altercation between a man and a police officer; he captured Officer Michael Slager shooting Walter Scott in the back, firing eight times as Scott tried to run away, and then, instead of rendering aid to the facedown, dying man, cuffing him.

Or they might think of Dylan Noble. A year after Scott was killed, bystanders captured footage of nineteen-year-old Dylan Noble lying on the ground in Fresno, California, as officers shot him multiple times. A month later, Alton Sterling's murder by police as he sold DVDs outside a Baton Rouge convenience store was captured on video by Abdullah Muflahi, the store's owner.[8]

While the nation was just learning about Sterling's killing in Louisiana, another man, Philando Castile—a well-liked employee in the Nutrition Services Department of the Saint Paul Public Schools—was shot by police during a traffic stop in Falcon Heights, Minnesota.[9] Diamond "Lavish" Reynolds, Castile's partner, was in the car when the officer shot him. As Castile lay bleeding in the driver's seat, the officer shouting curses outside, Reynolds began livestreaming the horror, calmly explaining to viewers what had just happened. Reynolds was handcuffed and, along with her young daughter who had been sitting in the backseat when the shooting occurred, was put in the back of the police cruiser; Castile died later that night.

Reynolds's video was shocking. The developers of Facebook's livestreaming capability probably had more lighthearted fare—like Chewbacca Mom—in mind when they developed the app. But Reynolds was confident in her decision:

> I wanted everyone in the world to know that no matter how much the police tamper with evidence, how much they stick together . . . I wanted to put it on Facebook and go viral so that the people could see. . . . I did it so that the world knows that these police are not here to protect and serve us. They are here to assassinate us. They are here to kill us because we are black.[10]

Reynolds's documentary impulse is widely shared. At Walter Scott's funeral, the pastor thanked God that Santana was able to capture Scott's death with his smartphone and bring his killer to justice. "Keep your phone handy, keep your charge up," he said. "You never know when you need to be around."[11]

Not all videos of police violence are filmed by chance. Many are the result of careful planning and vigilance by cop-watch groups. Alton Sterling's shooting was also captured on a smartphone by a member of a Baton Rouge nonprofit, Stop the Killing, whose volunteers had been monitoring police scanner traffic. Cop-watch groups have been around for decades, but the smartphone has been a game changer. Joaquin Cienfuegos, an organizer at Cop Watch LA, says livestream capability in particular has become a powerful new tool. Police often confiscate phones and erase video filmed by nearby witnesses. Livestreaming prevents this from happening, hindering cops' ability to create a false narrative.[12]

Various chapters of the American Civil Liberties Union, along with other groups, have even created free smartphone apps to facilitate easy filming and streaming. The ACLU of Michigan has released Mobile Justice MI, a smartphone app that allows citizens to "record, report, and witness" police interactions. "At a time when headlines and broadcasts abound with news about racial profiling, harassment, and police brutality," say the app's creators, "it is critical that citizens be equipped with the best tools available for ensuring law-enforcement accountability."[13]

The impulse to document police brutality is not confined to the United States. In Australia, oppressed minorities wage a similar struggle. Indigenous leaders in Perth, the capital of Western Australia, recently took part in a training workshop developed and facilitated by the National Justice Project, a nonprofit legal service based in Sydney. Aboriginal and Torres Strait Islander peoples are thirteen times as likely to be incarcerated as non-Indigenous Australians, and are subject to systematic harassment and abuse by police, particularly in Western Australia. The workshop focused on training Indigenous Australians to become citizen journalists, to use social media to highlight injustice.[14]

Highlighting injustice can be a dangerous endeavor, however. Stories pepper the web of people who've been intimidated, harassed, arrested, and beaten for filming interactions between law enforcement and civilians. In early 2019 US Immigrations and Customs Enforcement (ICE) officials raided a manufacturing plant in rural Sanford, North Carolina, and detained twenty-seven people. Sanford resident Christian Canales, a local musician who also worked the night shift at a nearby plant, livestreamed the scene, filming ICE agents as they checked the IDs of everyone leaving the factory. The Lee County Police Department was not amused; Canales was arrested and charged with "communicating threats" to law enforcement.[15] But these risks haven't stopped bystanders. Instead, over the past decade, the practice of livestreaming run-ins with law enforcement has spread.

These examples of how people are using smartphones to film the police aren't meant to capture the racism that permeates US institutions. Instead, they serve to open a conversation about technology and modern society—a conversation that moves from asking whether smartphones are having a negative or positive impact on society to one that asks how smartphones are being used to reflect and potentially reconfigure the oppressive relationship between law enforcement and people of color.

The ways this power dynamic are being reflected and reconfigured are fraught, to be sure. The emergent norm of filming the police as a strategy to hold them accountable or shape their behavior has changed the terrain of conflict in unexpected ways. One upshot is that police have responded by getting their own cameras. Police departments around the world began piloting body-worn cameras in earnest around fifteen years ago; in the UK police began wearing body cameras on domestic violence calls in 2006. But the public uproar over the killings of Trayvon Martin, Michael Brown, and Freddie Gray accelerated the adoption of body cameras by law enforcement, particularly in the United States. By 2016, 60 percent of local police departments and nearly half of sheriffs' offices had deployed body-worn cameras.[16] Both police departments and civil rights groups believed that body-cams would provide protection and accountability: cops thought cameras would protect and exonerate officers, while civil rights groups such as the ACLU believed cameras would reduce the use of force and increase police accountability.

The results of this rapid uptake have been contradictory at best. Prosecutors have leaped on using body-camera footage to prosecute civilians, while civilians often don't have access to the footage being used as evidence against them. Nor do ordinary people have a say in how police use their cameras, sparking fears over abuses related to facial recognition software and footage of vulnerable people shot without their consent, often in private spaces. More abstractly, the elevation of video as a primary tool for achieving justice (whether from the point of view of law enforcement or civilians) has moved the longstanding conflict between police and oppressed minorities onto a plane of dueling cameras. Police-worn body cameras show the cops' point of view while bystanders filming the same interaction with a smartphone often capture a scene that looks and feels quite different. The struggle to achieve justice becomes defined by whose footage a jury finds more compelling.

The newfound emphasis on filming also diminishes the salience of violent interactions that are *not* captured on film. In 2017, undercover police officers armed with AR-15 rifles shot up a car of unarmed teenagers in Hayward, California, killing sixteen-year-old Elena Mondragon. The incident exemplified gross police misconduct, yet because no one captured the incident on film, it received little coverage outside the local press. As Melissa Nold, a civil rights attorney assisting the family, told a reporter, "When there's no video, that's a battle for us. People just tend to believe what is reported by the police."[17]

These contradictions illustrate how technology embodies social relations, especially relations of power. In the struggle over police brutality, both those in power and those struggling to speak truth to power are attempting to harness a new technology to increase their leverage. The way this struggle is evolving demonstrates how our smartphones are simultaneously a window into the divides that shape American society and a tool being used to reconfigure those divides. This is an important duality: it not only sharpens our understanding of how smartphones fit into society, but also helps us to understand the evolution of other societal divides, namely the divide between men and women. The long and winding Black freedom struggle parallels and intersects with the women's liberation movement. In the past decade women have seized upon smartphones as a tool for agency and autonomy. But America's deep-rooted sexism is also being reproduced through our pocket computers.

Love Is on Our Phones

Love, particularly love on the small screen, is a major focal point in our smartphone society. There's Tinder and Bumble, the "feminist dating app" where women make the first move. There's Grindr, Zoe, and Transdr for the LGBTQ community, and Tin Dog for people who need their partner to love dogs as much as they do. Happn is for people who want to track down the attractive stranger they passed on the way to work, while Luxy helps rich people hook up with other rich people.

Finding love online is not exactly new. In his book *Dataclysm*, Christian Rudder, cofounder of OKCupid—a dating and social networking site that came online in 2004— boasts, "Tonight some 30,000 couples will have their first date because of OKCupid. Roughly 3,000 of them will end up together long-term."[18] Between 2005 and 2012, more than a third of Americans met their spouses online, and this number looks set to increase as attitudes toward online dating grow more positive.[19]

Dating apps have streamlined and "gamified" the online matchmaking models established by Match.com and OKCupid—that is, have added video-game-like features to increase participation, engagement, and loyalty. Tinder set the bar for the gamification of modern love, letting users sort and swipe potential lovers like an endless stack of playing cards. Happn has taken gamification to the next level with CrushTime: users get a notification when they've crossed paths with someone who has "liked" their profile and are presented with a four-profile grid of potential admirers. If they correctly guess which person tagged them, they win, "letting chance work its magic!"

We don't just find romance with our apps. We also manage it. Ghostbot sends tepid automated responses to unwanted suitors until they give up—a slightly more polite alternative to "ghosting," ending a personal relationship by suddenly and inexplicably cutting off all communication. Men concerned whether their member measures up can send a snap of it to "Eevie Bellini," a.k.a. The Banana_Butcher, on r/sexsells, Reddit's rapidly growing online smut marketplace. Bellini, one of many on the site offering similar services, writes three-paragraph evaluations for twenty dollars.[20] Women tired of fending off creeps or family matchmakers can purchase an invisible boyfriend, "a digital version of a real boyfriend without the baggage." For twenty-five bucks a month, invisibleboyfriend.com will send ladies photos, 200 text messages, and a personalized note, written by "real creative writers, not bots!"

Conglomerates such as Match Group, which owns Tinder, Match.com, OKCupid, and PlentyOfFish, are creating marketplaces for love (and sex), but they aren't the reason we're looking for love on the small screen. The emergent digital lovescape is rooted in much bigger shifts. The birth control pill, second-wave feminism, and the upsurge of women entering the waged workforce in the 1980s brought changes in attitudes about male and female roles. In the late seventies, roughly 60 percent of women agreed that "it is much better for everyone involved if the man is the achiever outside the home and the woman takes care of the home and family." By 2011, only 25 percent of women agreed. (Men were a bit more conservative, but their support dropped from 69 percent to 38 percent.)[21]

These shifts have accompanied evolving expectations and norms about marriage and relationships. Whereas once it was common to marry a neighbor or classmate and many women seemed more interested in escaping the confines of their parents' home than falling in love, by the 1980s nine out of ten women said they wouldn't marry unless they were in love. Today, sociologist and marriage expert Andrew Cherlin says marriage has become aspirational—a life step taken when "the rest of one's life is in order, including having access to decent paying, steady employment, which can take some time—or may never happen."[22]

In other words, adults from all walks of life are looking for love on Tinder, but better-off swipers are more likely to walk down the aisle. Most Americans eventually get married, but finding one's soul mate has become a long, drawn-out process that often includes numerous partners and long periods of cohabitation. Neither men nor women are in a hurry to get hitched. Dating apps facilitate a winding journey to love and marriage, offering access to a menu of partners unfathomable to our parents' generation.

Not everyone is enamored of modern love. Nancy Jo Sales, a best-selling author and contributing editor at *Vanity Fair*, wonders whether we're at the dawn of the "dating apocalypse." Sales paints a picture of feckless men who sport cracked iPhones and live with their parents, yet have so many women to choose from that they'll never need to settle down. They spend their days perfecting their "text game," swapping stories about "Tinderellas," and getting laid, while women, Sales contends, pretend that they're not hurt or dismayed by this behavior.[23]

The smartphone–social media combo has also introduced darker elements into the twenty-first-century lovescape. Cyberspace, like regular space, is ripe with misogyny, sexual harassment, and threats of violence. Women searching for love on the small screen are often greeted with unsolicited dick pics and aggressive sexual requests. Digital networks can provide crucial support for women trying to escape abusive relationships, but they can also make it harder for women to shake violent partners, who often have access to their digital networks and, increasingly, their smartphones. A 2014 study by the National Network to End Domestic Violence found that more than half of domestic abusers tracked their victims' smartphone activity (call logs, text and chat messages, GPS data, photos) using stalkerware apps such as mSpy and FlexiSpy.[24]

Despite these downsides, a broader pattern is clear in the United States: women's choices—particularly those of well-off, educated, white women—are increasing; women have more agency and autonomy and are actively using their smartphones to exercise choice and pursue happiness in love.

The situation is quite different, however, for American teenagers, particularly teenage girls. Most adults have accepted the fact that teenagers between the ages of fifteen and nineteen are having sex (even though both teens and adults today appear to be having less sex than previous generations), but they are *not* OK with teens sending naked or risqué pictures of themselves to each other. Parents and policymakers are, to put it mildly, horrified by sexting. Periodic newspaper exposés, like the one about Cañon City

High School in Cañon City, Colorado, confirm their worst fears. Cañon City high schoolers were caught with hundreds of pictures of naked girls on their phones.[25] The male students, many of whom were on the football team, had made a game of collecting and sharing photos of girls. Students hid the photos using "vault" apps they'd downloaded to their phones; the mobile apps are disguised as an ordinary calculator app, but actually provide password-protected storage for photos and videos.

The Cañon City case is not unusual. Similar stories abound. In case after case, adults are shocked while teens often appear nonchalant about their activities, offering an unconcerned shrug and saying "Everyone is doing it." Parents, law enforcement, and school administrators find this response exasperating and in recent years have made moves to implement stricter regulation of sexting. Sexting by minors, even to another minor, is a felony; and since 2009, many states have ramped up punishment for sexting. The US House of Representatives passed a bill that could send teens who sext to federal prison for fifteen years.[26] The argument is straightforward: law-makers say sexting puts teens at risk of sexual predators and promotes child pornography.

There has been some pushback against equating consenting, same-age teens with sexual predators, and prosecutors often decline to prosecute high school students involved in naked picture rings. Some experts even advocate the right of teens to sext, arguing that it's a healthy and relatively safe way to explore sex. But the antisexting consensus remains firm. The vast majority of parents are not even a little bit OK with sexting. They're baffled as to why teens are, in their eyes, so reckless and unable to foresee how upset they'll be if a racy picture of them gets posted on social media.

The concern is not directed equally at boys and girls, however. For the most part, warnings about the dire consequences of sexting are directed squarely at teenage girls. Girls who send naked pictures of themselves to boys (or other girls) are depicted as suffering from low self-esteem, mind-lessly recreating the sexualized images they consume on a regular basis, or weakly giving in to biological urges. If they sext a boy, it is because they are caving in to the boy's demands. Yet, while girls are presented as victims of bi-ology, psychology, or mass media, they are simultaneously expected to con-trol and prevent sexting. When naked pictures of a girl are passed around without the girl's permission the girl is usually the one who gets blamed because she made the wrong choice to send the picture in the first place.[27]

The End of Men author Hanna Rosin's story about a sexting scandal in Louisa County, Virginia, shows this dynamic in action. "Briana," a student at the local high school, was one of a number of girls whose naked pictures had appeared on an Instagram account. Briana had sent the picture to an older boy who'd relentlessly badgered her for it, only to ignore her after she sent it. The boy made the picture public, but Briana was the one punished. Classmates blamed Briana for the investigation that ensued at the high school, and shunned and mocked her. She lost her babysitting job after the mother of her charge would no longer speak to her. Briana's own mother struggled to get past the incident, warning Briana's little sister, "Don't end up like your sister!"[28]

The age-old double standard is rebooted for the smartphone age. As Peggy Orenstein argues in *Girls & Sex*, girls are either "hoes" if they sext or "prudes" if they do not—a negative either way. Boys who engage in sexting are perceived to be engaging in a bit of harmless sexual experimentation, or better yet, are seen as "players."[29] The end result, contends sociologist Amy Adele Hasinoff, is that predatory men and boys are positioned "as an inevitable and natural part of the cultural landscape." Girls are blamed "for online victimization rather than those who violate privacy and harass others." Girls are taught that their self-respect is contingent upon "their ability to exercise caution and self-censorship online."[30]

Whether we think teens should or should not be sending naked pictures of themselves to each other, our phones reveal how, amid a broader moment of rapidly shifting norms about relationships, love, and sex, patriarchal tropes are being reinforced and reconfigured in messages aimed at young women and men. For adult women, the smartphone is a tool of liberation, its use couched in practices that support the broadening of femininity, challenge heteronormativity, and reinforce the importance of happiness and choice. For young people, however, the boundaries of modern love are drawn in ways that often perpetuate patriarchy. Adult policing of teen smartphone use, in particular teen sexting, while understandable within longstanding cultural frames, reinforces the sexist norms that underlie rape culture and victim blaming.

These glimpses into modern love and documenting police violence highlight how the ways we *use* our hand machines reflect and reconfigure persistent divides—sexism and racism—in American society. Other divides are built into the architecture of smartphones themselves.

Built-In Divides

Apple insiders refer to FoxConn's Shenzhen factory as Mordor, Tolkien's Middle Earth hell hole in *The Lord of the Rings*. As a spate of suicides in 2010 tragically revealed, the moniker is only a slight exaggeration of the conditions in factories where young Chinese workers assemble iPhones. The pristine white box holding our shiny new device hides this messy reality. We can't see the relationships, and divides, built into smartphones. But they are there all the same.

The architecture of our phones is a topological map of global inequality. Apple's supply chain links colonies of software engineers with hundreds of component suppliers in North America, Europe, and East Asia: Gorilla Glass from Kentucky, motion coprocessors from the Netherlands, camera chips from Taiwan, and transmit modules from Costa Rica funnel into dozens of assembly plants in China where workers toil in stressful, low-paid jobs.[31]

Power and governance are located at multiple points in the smartphone value chain, and production and design are deeply integrated at the global scale.[32] But the configurations of power that begin with mining the rare metals needed to produce smartphones—lithium in Bolivia, coltan in the Democratic Republic of Congo where miners, many of them children, dig by hand—and end with Apple and Google, reinforce existing wealth hierarchies. Poor and middle-income countries try desperately to move into more lucrative nodes through infrastructure development and trade deals, but upgrading opportunities are few and far between. The global architecture of production and trade is designed for the benefit of multinational corporations and wealthy countries, making struggles by workers to improve conditions and wages extremely difficult.[33]

Yet, the men and women and children who do the dirty, dangerous jobs that make our smartphone reality possible—the miners, the assemblers and dis-assemblers—are not who come to mind when we think about the connection between people and smartphones.[34] As literary critic Frederic Jameson observes, "The truth of metropolitan existence is not visible in the daily life of the metropolis itself." We can't see the global divides that make the consumption patterns of wealthy countries possible.[35] Instead we imagine the smartphone user, the consumer who can endlessly and effortlessly stream movies, Snapchat friends, shop the web, find the museum.

The divide between those who make smartphones and those who consume them, particularly Western consumers, is just one element of what experts call the "digital divide."[36] As socioeconomic infrastructure becomes more and more dependent on digital connections the gulf between those who have easy access to high-speed internet and those who don't appears starker than ever. In the United States, at least 24 million people, including a quarter of rural Americans and 5 million households with children, have no access to broadband internet service.[37] In Cambridge, Massachusetts, a city that houses both Harvard University and the Massachusetts Institute of Technology, 40 percent of low-income households lack broadband access.[38]

In the late nineties, telecom providers such as AT&T and Comcast, eager to gain market share, invested heavily in expanding their networks. But this expansion has proved to be highly skewed. The Haas Institute reports that in California AT&T has deployed its most advanced broadband technology in high-income communities, while more than four million households in the state lack access to high-speed broadband.[39] AT&T has also been accused of "digital redlining" in Cleveland, Ohio. A report by the National Digital Inclusion Alliance and Connect Your Community, a Cleveland nonprofit working to overcome "digital exclusion," found that AT&T withheld broadband improvements in Cleveland's poorest neighborhoods.[40] Broadband providers and their shareholders don't want to spend money connecting areas where revenue streams might be lower or costs higher. But telecom companies don't want other providers to muscle in on their territory, either, so they have lobbied dozens of states to pass laws prohibiting rural areas and certain urban clusters from building their own municipal fiber-optic networks.[41]

Caught in this bind, rural and urban poor folks rely heavily, and often exclusively, on their phones for access to the internet. For at least 20 percent of Americans their smartphones are their only lifeline to the digital world. But this lifeline has limited capability. It's difficult or impossible to apply for a job, conduct research for a class project, or craft a college essay on a smartphone. Phone plans are also expensive and minutes run out, leaving users who have no home connection looking for Wi-Fi hot spots. The name for these seekers is "leaners"—people who lean against buildings trying to pick up Wi-Fi, until the police make them move on. The term calls to mind Jean-François Millet's famous painting *The Gleaners*, which depicts three young women gathering leftover stalks of wheat after the harvest.

Elites and highly paid professionals live in a different world from leaners. For them, the smartphone dream is a reality. With a few taps and swipes they can restock milk, eggs, and bread from Instacart; avoid the lunch-time rush with a tasty bite delivered through Seamless; hail a Lyft when after-work cocktails turn tipsy; or hire a "tasker" from TaskRabbit to do a deep clean of the condo before the in-laws arrive. The smartphone brings convenience, a prime commodity for busy professionals who struggle with what sociologist Arlie Hochschild calls the time bind.[42]

Until recently, women took care of pretty much all the household chores and childrearing duties, even after they entered the waged workforce en masse in the 1980s. But these norms are slowly changing. As both women and men work long hours men are picking up (slightly) more of the slack, particularly as pressure to engage in a "concerted cultivation" philosophy of childrearing permeates the middle class.[43] With smartphone apps such as TaskRabbit and Amazon Shopping, which now offers personal services, professional couples don't have to squabble over who's going to go to the grocery store or pick up the dry cleaning. They can outsource these chores to someone else. They can afford to. The top 10 percent of the American population has seen its income rise considerably in recent decades. Money buys convenience.

Wealthy people have been hiring poor people to scrub their toilets and mow their lawns for a long, long time, but something is different today. Norms have shifted. Apps, and the positive ideas associated with them, have erased the stigma of hiring someone to assemble your IKEA haul, walk your dog, or wait in line for you at the new brunch place. It's like *Downton Abbey*, the wildly popular BBC show about the trials and tribulations of a rich, late–Belle Époque English family and their servants, except the hired help don't live in your house. Instead of ringing a bell for service, you tap an app. A maid, masseuse, taxi driver, personal shopper, whatever you want—"there's an app for that."

It's not entirely clear how many people are working in these app jobs—piecework gigs mediated through a smartphone app. In the United States, estimates of "on-demand" app workers who earn money via online intermediaries such as TaskRabbit, Lyft, Uber, and Amazon Shopping vary widely. The Federal Reserve estimated that in 2017, 16 percent of adults earned money from app jobs, while the Bureau of Labor Statistics reported that 3.8 percent of workers (5.9 million people) were classified as contingent workers in 2017.[44] Even absent concrete numbers, it is clear that the emergence of app jobs in the past decade is a significant development in the evolution of work in the United States.

App jobs are the next step in what the employment and labor market expert David Weil calls the "fissuring" of the workplace, whereby employers "change the boundaries of the firm itself": "Employment is no longer the clear relationship between a well-defined employer and a worker."[45] The spread of multitier subcontracting has enabled companies to plan their budgets around paying for services rather than paying wages. App jobs are a testing ground for employers to see how far they can go in externalizing both work and workers, to see how much they can force workers, consumers, and governments to take on the costs of production.

Uber, the largest provider of app jobs in the United States, shows this externalization at work. Drivers provide their own vehicle (paying for gas, repair costs, and insurance) and phone, while Uber provides only the software and its network. Uber takes a 25 percent cut from each ride, yet drivers are not considered employees and the company is not responsible for their safety. Uber is also not responsible for the safety of its consumers (riders) nor the increased congestion and pollution it causes in urban centers. In San Francisco, for example, transportation experts concluded that transportation network companies such as Uber and Lyft were responsible for more than half of the increase in roadway congestion between 2010 and 2016.[46]

In trying to grasp the scale of smartphone jobs, we must also include all the warehouse and logistics workers continually set in motion by our "wherever, whenever" purchasing power. We order a last-minute birthday gift for Mom on the train to work; we re-up the laundry detergent during a commercial break; we order another pair of earbuds on the taxi ride home upon realizing that we've left ours on the plane. During the busy shopping season after Thanksgiving in 2017, Amazon US saw a 50 percent spike in mobile shopping—shopping done on mobile devices—over the previous year. The company says mobile shopping now accounts for more than 40 percent of its revenue and it's growing by leaps and bounds.[47]

Every time we impulsively tap a purchase on our phones, someone is on the other end, filling a box with whatever we ordered before handing it to someone else to drop it on our doorstep. Amazon alone employs more than 613,000 warehouse workers worldwide, and adds about 100,000 more temp workers during peaks.[48] Jessica Bruder, in *Nomadland: Surviving America in the Twenty-First Century*, follows the lives of Amazon's "CamperForce," a large group of (mainly) retirees who can't afford to retire, so they live in their RVs and other vehicles and find temporary work in the warehouses of "the everything store" during the holidays. When peak season is over, these "workampers" drive away in what Amazon executives proudly call a "tail light parade."[49]

Log on to the Spare5 mobile task app on the long bus ride home, circle the road signs in a series of photographs for some self-driving-car start-up, and you've paid for that afternoon splurge on a triple macchiato at Starbucks. App jobs like this are appealing at first glance. There are few barriers to entry. You can work when you're able, and the ads promise decent money. You just need a smartphone, often a car, and a willingness to work. It certainly seems better than the frustrating world of scheduling software and low-paid, irregular shifts common in the retail and fast-food sectors.

In some respects, it feels like being your own boss. Tech companies push this interpretation, emphasizing how they provide the digital platform—digital infrastructure that facilitates interactions (often commercial) between at least two people or groups—and you provide the hustle. Uber, for example, is adamant that it is not the employer of the roughly one million drivers worldwide who use its app to find people to ferry around. Researchers at Carnegie Mellon's Human-Computer Interaction Institute aren't so sure. They call the Uber arrangement "algorithmic management."[50] App workers who use these platforms to earn money don't have easy access to a flesh-and-blood manager. Instead they interact with an algorithm—a set of exact instructions to solve a problem or perform a computation. Algorithms can be written to perform simple tasks, like adding or subtracting numbers, or complex tasks, such as playing a video or, in the case of Uber, telling the driver where to drive, paying them what they are owed, and so forth. Algorithms effectively transform Uber drivers' phones into their boss.

Turns out, it's not so great to have a smartphone as your boss. The modern work relationships enacted through our phones show not only how phones are a dream come true for the well-off, but also how the twenty-first-century working class is being made, and the divide between the haves and the have-nots is being reinforced.

In the United Kingdom, drivers for Deliveroo, a food delivery app—disproportionately immigrants and poor people of color—are closely monitored by Deliveroo's algorithm. They have thirty seconds to respond to the app when it pings them, and they don't know where they're going until they swipe "accept delivery." If they don't accept they get punished. They also get punished for being too slow; if the driver's "time to accept orders," "travel time to restaurant," "travel time to customer," or "time at customer" are longer than what the algorithm estimates they should be, the driver's account can be deactivated.[51]

Uber drivers have even less time to respond to "trip requests," ten to twenty seconds, and they also don't know where they're going until they've picked up the passenger. Until recently drivers got time-outs—short periods where they are locked out of the app—if they refused three trip requests in a row. Reliable data on how much pay Uber and Lyft drivers take home is hard to come by, but a recent driver-earnings survey found that drivers of Uber's most popular service, Uber-X, made a median wage of $14.73 an hour in 2018 after tips but before gas, insurance, and repairs—substantially less than a living wage.[52]

Amazon warehouse workers are algorithmically managed in a different way. Each "picker" has a GPS monitor that tells her precisely which way to walk to get to the product she's looking for and the number of seconds it should take her to get there. If she walks a different way, or takes too long, she'll get a warning and possibly a demerit, and too many demerits can add up to dismissal.[53] Amazon's model exemplifies the steady intensification of work over the past few decades. Fifteen years ago, *New York Times* columnist Thomas Friedman marveled over Walmart's use of computerized headsets to direct its warehouse forklift drivers.[54] Today, truck drivers, nurses, office workers, plumbers, technicians, salespeople, and more are tracked by their employers through GPS-enabled apps on their phones. Many don't even know they are being tracked, because employers are not required to tell them, and a significant percentage are tracked twenty-four hours a day, whether they're working or not.[55]

Moreover, many algorithmically managed workers don't have the rights that regular employees have. Those designated as independent contractors—pretty much all on-demand workers today—are not entitled to minimum wage, sick days, overtime pay, safety protections, unemployment or health insurance, a pension plan, or disability pay. App workers in service jobs are held hostage to the reviews of fickle customers, and if they have a problem, their app boss isn't much help. Melissa, an app driver for the food delivery service DoorDash, voiced a common complaint in an article that appeared in the *Guardian* in 2017. A customer sent Melissa a pornographic video through the DoorDash app, but when Melissa tried to report the incident it was extremely difficult to reach an actual person at the company. She finally reached someone at DoorDash only to be told there wasn't much the company could do, and that as an independent contractor she should know the risks. Melissa was on her own. When a customer grabbed her breast a month later, she didn't even bother reporting it.[56]

Some app workers say they really like what they're doing and appreciate the opportunity to earn some extra cash. That may be so. But it's also important to be clear-eyed about the types of employment relationships being generated through our smartphones. Environmental activists use the term "greenwashing" to describe how companies use superficial tricks and PR spin to hide environmentally destructive behavior and present a green, responsible image.[57] Tech companies engage in "appwashing," using slick apps and fuzzy stories to hide the reality of the low-paid, stressful jobs they create and depend on for their profits.

Amazon's 2017 holiday commercial is a perfect example of appwashing. It shows a woman on a bus getting a goofy Snapchat from (presumably) her niece. Inspired, she instantly taps her phone app to send the little girl a gift, brought to life as an adorable Amazon box singing Supertramp's "Give a Little Bit." The happy box makes an epic journey, singing all the way, through a maze of conveyer belts, under stamping machines, on trucks and planes, before finally being handed to a smiling child thousands of miles away. It's a logistical miracle. But the human workers who make it all possible are obscured or nonexistent. We see sparse, shadowy forms, blurred faces, glimpses of hands. The only people depicted clearly in the commercial are the woman who ordered the gift with a tap of her finger and the happy little girl who receives it. This is the unconscious message that we absorb from our phones today: that the people who make our emergent app economy possible are interchangeable, invisible, or unimportant.[58]

A Bigger Conversation

Talking about mobile justice apps, sexting shame, digital divides, and appwashing starts a bigger conversation, one that moves beyond debates about whether smartphones are helping or harming society to a more dynamic discussion about power in America. The stripped-down snapshots we just saw of how race, gender, and class are being reproduced and reconfigured through our smartphones encourage us to put our phones at the center of this bigger conversation. As the French philosopher Gilles Deleuze said, machines "express those social forms capable of producing them and making use of them."[59] Looking at the ways we interact with our smartphones reveals the social relationships and power structures that undergird modern society.

This closer look at our pocket computers demonstrates how they have been quickly repurposed as a tool to chip away at long-standing divides, such as the divide between Black people and the American police state. As Diamond Reynolds articulated so clearly when her partner, Philando Castile, was murdered, Black Americans are using their phones to document the continuation of centuries of racism and state violence. Smartphones are not a solution to racist policing. The failure of body cams and damning footage, such as that streamed by Reynolds, to bring justice most of the time reiterates that there is no technological fix for white supremacy. Instead, smartphones have been woven into an ongoing political struggle in ways that both reflect and reconfigure that struggle.

Smartphones are also playing a role in the long arc of women's emancipation from the domestic sphere. Women, alongside men, have enthusiastically taken up smartphones as a device to exercise choice and pursue happiness in love. Yet, comparing the experiences of women and teenage girls in exploring love on the small screen illustrates the powerful hold of patriarchy. While adults are right to be concerned about sexting, a clear double standard exists: When teenage girls use their phones to explore their sexuality they are censured and shamed; teenage boys are given a pass for the same behavior, and are rarely punished for violating privacy and trust. This disparity reinforces the divide between men and women, reassuring young men of their power and rights and young women of their inferiority and deficiencies.

A conversation about power and smartphone technology also makes clear how our phones encapsulate the divide between the rich and the poor; the socioeconomic chasm between the workers who mine the coltan and assemble the chips and the downstream consumer, is embedded in the architecture of our phones. We don't see these social relationships when we peer down at our phone screens. The making of the twenty-first-century working class is also obscured through mechanisms such as appwashing. The companies that have transformed smartphones into unimaginable tools of convenience rely on a steady supply of folks desperate enough to declutter the closets of Brooklyn's elite, walk miles in a stifling warehouse packing boxes, or pedal through midday traffic to deliver a BLT to a Boston lawyer. Smartphone-mediated conveniences rely on a system of stressful, degrading, low-paid work that is hidden behind cute apps and feel-good stories.

Highlighting how our phones reflect and reconfigure longstanding divides over race, gender, and class isn't where the conversation about power and technology ends—it's where it begins. By seeing our phones as both a tool—used not just by people but also by companies, governments, community groups, and political movements—and as an embodiment of unequal power relationships, we can begin to understand the shifts occurring around us. We can begin to comprehend how power is being reconfigured, consolidated, and challenged in our smartphone society. That's where we're headed in the rest of this book.

Chapter 2

New Titans

A 1938 promotional video for the Ford Motor Company's River Rouge complex in Dearborn, Michigan, painted a picture of power and vision: the Rouge sprawled over 1,096 acres, the film boasted, with 7,250,000 square feet of floor space, 345 acres of glass windows, 90 miles of railroad tracks, 3 towering blast furnaces, 220 ovens for making steel, and a 400-foot-long glass-baking oven, altogether consuming as much electricity as the city of Cincinnati. Ford's steamers would make their way along the Red River carrying coal, limestone, ore, and other raw materials to the sprawling industrial city where 80,000 workers would transform these raw materials into finished automobiles, made to the exact specifications of "Mr. and Mrs. Customer." It was an audacious vision of mastery over man and machine. "Methods and machines change. The earth and man endure," the video's narrator intoned; Henry Ford was a changemaker, a titan of industry near the end of a long line of such industrial titans.[1]

Morgan, Rockefeller, Carnegie, Vanderbilt. The railroads, steel mills, and financial behemoths built by these captains of industry transformed American life in the late nineteenth century. They pushed families off the farm to the cities, drew millions of immigrants to the country's shores, and established the United States as a "great power." The mammoth concerns also made these men incredibly rich. Andrew Carnegie, who headed the world's first billion-dollar corporation, was a "magician of money," according to historian Jackson Lears. One the eve of the Great War, John D. Rockefeller was worth nearly a billion dollars, about $190 billion in today's dollars. In 1895, the banker J. P. Morgan bailed out the US government when it ran out of gold.[2] Historian H. W. Brands writes, "Never had a class of Americans been so wealthy and never had such a small class wielded such incommensurate power."[3]

These men and the companies they built are no longer titans. Some of them, like the Ford Motor Company, are still around and are still powerful. But the companies and visions these men built no longer animate our imaginations or loom large in the public discourse. Most often these companies are invoked as examples of a dynamic and prosperous past. Today, new titans are transforming American life. You'll find their icons on your smartphone screen.

The top ten apps people kept on their smartphones in 2017 were the following:

1. Facebook
2. Gmail
3. Google Maps
4. Amazon
5. Facebook Messenger
6. YouTube
7. Google Search
8. Google Play Store
9. Instagram
10. Apple App Store

Mark Zuckerberg, Sergei Brin, Larry Page, Jeff Bezos—these are the titans of today. Facebook is social media. Google is search. Amazon is e-commerce. Jeff Bezos, founder and CEO of Amazon.com, is worth $112 billion. Mark Zuckerberg, the CEO of Facebook, is worth $69 billion. Larry Page and Sergey Brin, the founders of Google, more than $50 billion each. But these men's titanic profiles are based on more than their wealth. They have created institutions that are integral to modern political, social, and economic life. When we pull out our phones to connect with the world, the vast majority of us use architecture created by these companies. And while our pocket computers are much smaller than the machines that powered the Rouge, the vision of mastery behind the tech titans who dominate our smartphones is no less audacious.

Masters of Cyberspace

Facebook has more than two billion active users who on average spend fifty minutes a day on Facebook and other Facebook-owned apps, Instagram, and

WhatsApp. This is nearly 80 percent of mobile social media traffic. In many countries, says Cynthia Wong of Human Rights Watch, "Facebook is the de facto public square."[4] Google is the world's search engine (outside of Russia and China), controlling over 80 percent of the world search engine market and over 90 percent in the United States and Europe. It processes 1.2 trillion search queries a year. Meanwhile, at YouTube, Google's video platform, the equivalent of Hollywood's entire catalog is uploaded every few days. Eighty-three percent of all new ad dollars worldwide are spent at Google and Facebook. Nearly half of all e-commerce flows through Amazon, and in 2017, more than half of all product searches began on Amazon. Amazon also controls a third of the market for cloud infrastructure—the hardware, software, and facilities necessary for cloud computing.

Our phones are central to the meteoric rise of the tech titans. "Mobile tech acted like rocket fuel for the Internet," transforming it from something millions of people participated in to something billions participated in.[5] Today, roughly 82 percent of smartphones worldwide run on Google's Android operating system and Google controls 97 percent of the mobile search market. Three-quarters of the people who use Facebook do so from their phones. When they watch YouTube, more than half are streaming through a smartphone. WhatsApp (with 1.5 billion users), Instagram, (1 billion users), and many more "killer" apps were designed just for phones. Smartphones have overtaken computers as the place where people buy things from the web. Boston Consulting Group data shows that the mobile ecosystem generated 4.2 percent of global GDP in 2015, about $3 trillion. Experts say that these days, Silicon Valley is firmly "mobile first" in its development and design focus.[6]

As we were with the titans of old, we're a bit obsessed with these new companies and the (mainly) men who run them, treating them with an admixture of reverence and disdain. Once we shared stories of Vanderbilt's lavish parties and Morgan's black yacht that he pretended was a pirate ship. Now we gossip about Bezos's "midlife crisis physique," Zuckerberg's penchant for gray T-shirts, and Page's possibly vampiric quest for the fountain of youth.

Only recently, however, has American society begun to come to terms with the power these new companies hold. Posing as earnest geeks who just wanted to give us cool stuff, these new titans snuck up on us. Compared to the Wall Street "masters of the universe" who nearly brought down the global economy in 2008, these new companies seemed to have bigger, more interesting things in mind than making a quick buck. They do. But they're also pursuing these things with a ruthlessness reminiscent of the robber barons of yesteryear. As it becomes clearer and clearer that the titans of today are reorganizing the world through our pocket computers, it is imperative that we know what their true aims are.

Humans were really excited about the internet twenty years ago. We told stories about the freedom, democracy, and sharing it would bring. Visions of individuals meeting other individuals on a free and equal internet plane were mesmerizing.

These stories were from the start mostly fanciful. The internet was created and developed by the United States Department of Defense. War-making and surveillance are written into its DNA. Moreover, in the early nineties, when the internet finally looked like it could be profitable, after decades of taxpayer-funded research and development, the US government quietly gave it away to a handful of private telecom providers—a move some deem the largest financial giveaway in American history.[7]

Nonetheless, a little bit of the internet fairy dust was real. The World Wide Web's creator was Tim Berners-Lee, a man who saw the internet as a space where knowledge and information could be collectively created, owned, and shared—a digital commons. Anyone could create a webpage to share musings and ideas with the world, and thanks to net neutrality, everyone could see these sites if they were connected to the web. Unlike radio, newspapers, and television, the internet promised to be an online space that wasn't controlled by big corporations.

It's difficult to remember those early days. The internet is a lot prettier now; the clunky blogs, unslick online sellers, and search engines that didn't find things very well have become the stuff of nostalgia. Cyberspace has been colonized by a few corporations who've brought their ethos of clean, crisp, and fast to everything, including and especially our phones: Facebook's simple wall to chat with friends; Google's uncluttered home page waiting for your query; Amazon's promise to have any book, and soon anything, you wanted.

Today, these companies have moved far beyond their humble home worlds. They have created whole "ecosystems" that dominate social media, online search, e-commerce, and cloud computing. They are also so much more. Growing at a rapid clip, largely by swallowing up smaller rivals or companies whose capabilities would augment their own, Facebook, Amazon, and Google have come to dominate digital life.

Amazon began with an ambition to be *the* online bookseller; today it controls 72 percent of the online book market and sells millions of other products in its online marketplace. As it branched out to become the "everything store" it purchased Zappos, Diapers.com, Twitch, and Audible. Recently, Amazon got into the business of lending money, originating $3 billion in loans to small sellers who use its "marketplace" platform. In 2017 it bought Whole Foods for $13.7 billion, and in early 2018 it signed a joint-venture deal with investment guru Warren Buffett and J. P. Morgan to reconfigure the health-care market. This is on top of Amazon's wildly successful venture Amazon Web Services (AWS), which provides on-demand, pay-as-you-go cloud-computing platforms for individuals, small and large businesses, and even governments, and generates half of the company's profits.[8] AWS customers include Netflix, CapitalOne, Condé Nast, and the Central Intelligence Agency, which paid Amazon $600 million for cloud space.

Google has developed an equally voracious appetite since its founding. Google owns YouTube and Android and, of course, Google Search. It is also the key piece of a sprawling conglomerate called Alphabet. Google founders Sergey Brin and Larry Page created Alphabet in 2015 to organize their growing pile of tech companies—a conglomerate that, in addition to Google, includes companies focused on biotech (Calico), cybersecurity (Chronicle), wind power (Makani), and the life sciences (Verily). Add to that Waymo and Wing, which develop self-driving car and drone delivery technology, and DeepMind, Alphabet's artificial intelligence subsidiary. Throw in venture capital and private equity firms (GV, Capital G), a tech incubator (Jigsaw), broadband and balloon internet providers (Google Fiber and Loon), an urban innovation organization (Sidewalk Labs), and a "semisecret" research and development facility called X Development, and one gets a sense of the growing reach of the behemoth that started with a search engine.

Facebook, too, has made big-ticket purchases such as Instagram, WhatsApp, and Oculus, and when it can't buy what it wants, as in the case of Snapchat, it has implemented similar features on its own platform. Just as Google is the world's search engine, Facebook wants to be the world's social media platform. To overcome the hurdles of poor populations with little or no access to the internet, Facebook created Free Basics (now called Internet .org), a program to provide subscribers who can't afford either a broadband connection or a smartphone data plan with a free, bare-bones version of the internet. The idea is that minimizing data transmission limits costs. Of course the bare-bones version includes Facebook.

The limits of the plan were highly criticized in some countries; in India opponents argued that Free Basics positioned Facebook as a gatekeeper threatening net neutrality.[9] But where the plan has been adopted, Facebook has garnered millions of new users. For example, in Myanmar, Facebook's user base increased from 2 million in 2014 to 30 million in 2017.[10] To build its user base even further, Facebook has been secretly working on a satellite called Athena, which it plans to launch into space to beam down internet to rural regions with no or limited internet connectivity using high-frequency millimeter-wave radio signals.[11]

Of course these aren't the only tech titans. Apple, under the leadership of the late Steve Jobs, created the smartphone, setting in motion many of the dynamics we associate with our smartphone society. Apple vies for the title of world's most valuable company and in September 2019 was worth more than $1 trillion. Its iPhone captures roughly 11 percent of the world smartphone market and is a key trendsetter in global phone and computer technology, particularly on the hardware side. Microsoft controls 17 percent of the cloud-storage market, continues to be a major software provider for business and institutions, and recently joined Apple and Amazon in hitting a $1 trillion valuation. Ninety percent of the US wireless market is controlled by four telecom companies, of which the biggest two, AT&T and Verizon, together control more than 60 percent of the market.[12] Meanwhile, companies such as Uber, Airbnb, and Twitter have created widely used apps that are disrupting existing markets and creating new ones.

Other countries have their own titans. China has Alibaba, started by Jack Ma in 1999 in his Hangzhou apartment. Today Alibaba is among the world's ten biggest companies by market capitalization and can count a sprawling online marketplace, digital video, music, sports, an online payments system (called Ant Financial), and artificial intelligence research in its portfolio. Baidu and Tencent are Alibaba's peers. Their stakes in ride-hailing apps, media, food delivery, and e-commerce have made China the world leader in mobile payments. According to the McKinsey Global Institute, one-third of the world's 262 "unicorns"—start-ups valued at more than $1 billion—are Chinese.[13]

Tech titans are not the only large and in-charge companies. The US and global landscape has come to be dominated by "superstar firms" in nearly every major sector: manufacturing, retail, finance, services, wholesale, utilities, and transportation.[14] In each of these, one or a few corporate players dominate and take the lion's share of profits. This isn't surprising, really.

The tendency for individual business owners and corporations to increase the amount of capital they control, and for individual concerns to be combined into new, larger agglomerations, is a persistent dynamic in capitalism. But until a few decades ago, there was resistance in the United States to letting companies get as big and powerful as Facebook, Google, and Amazon are today. The financial shenanigans, insider trading, violence, and price hikes that accompanied the rise of the Gilded Age titans, combined with the dogged campaigning of early-twentieth-century Progressive advocates for people-centered corporate reform, had lasting influence on policymakers and the general public.

Until the 1970s the United States had a fairly aggressive set of antitrust laws and practices designed to foster fair competition between firms and by extension to protect consumers from predatory business practices. Creating space for small businesses to grow and prosper was valued, even if only rhetorically. Granted, the nature and practices of antitrust regulation varied over time. AT&T was given monopoly rights in its early years but was tightly regulated by the US government, which used tax dollars to help AT&T fund and develop research.[15] On the whole, however, there was a bipartisan consensus that too-powerful companies should be broken up for the good of consumers and capitalism as a system.

The triple crisis (economic, social, political) of the 1970s catalyzed a broad shift in ideas about how to regulate and grow capitalism.[16] The New Deal assumption that the government should take an interventionist role to protect stakeholders from business overreach and malfunctioning markets lost sway, and a new legitimating framework emphasizing shareholder value, efficient markets, and deregulation came to the fore. Ronald Reagan epitomized these values, promising to get the government "off the backs of the people." Companies in all sectors began to get a lot more breathing room through tax breaks, decreased oversight, watered-down environmental laws, and defunded federal regulatory agencies. At the same time, corporate mergers began to be viewed with a much less skeptical eye.[17]

There have been exceptions of course. Microsoft got on the bad side of the Clinton administration, got sued by the Department of Justice, and was forced to tweak its business model to make its software friendlier to competitors by sharing its application programming interfaces with third parties. But the case wasn't a trendsetter. When Federal Trade Commission staffers advocated bringing a similar lawsuit against Google in 2012, the FTC commissioners quickly overruled them.[18] Jeremy Stoppelman, the CEO of Yelp (which competes with Google in providing ratings and reviews of local services), has long argued that Google abuses its power in search; Stoppelman insists that Obama regulators were uninterested in his warnings.

Until quite recently, the word "monopoly" has not been negatively associated with tech. On the contrary, Peter Thiel, the cofounder of PayPal, argued in *Zero to One* that competition is bad for capitalism—or at least, for profits. "Creative monopoly" is something to aspire to, something that brings success and innovation.[19] A few years ago it would have been perceived as absurd to suggest that Facebook or Amazon should be regulated in a similar way to Vanderbilt's railroads or Morgan's steel. Apples and oranges.

Part of the reason why it seemed absurd is that today's titans have built companies that look quite different from old-style corporations—they've built digital platforms that connect groups together, allowing them to interact, rather than focusing on selling a particular product or service. Ford sells cars; Facebook connects people who love cupcakes with other people who love cupcakes. The upshot is a broader transformation in how goods and services are produced, shared, and delivered. Instead of the tired conventional model, with individual firms competing for customers, platform enthusiasts say we are witnessing the emergence of a new, seemingly "flatter," more participatory model whereby customers engage directly with each other through an architecture designed for connection.

This focus on connection, combined with our *ease* of connection, thanks to our smartphones, has generated unprecedented "network effects": the more people connected to a network or platform, the more powerful, useful, and profitable that network or platform will become. The result is the formation of "natural" and essentially harmless monopolies. According to this line of argument, if we try to regulate these (harmless) monopolies, we'll destroy the network effects. We'll kill the goose.[20]

Besides, who cares if these companies are dominant? They are for the people! Jeff Bezos, Amazon's founder and chief executive, put his shareholders on notice in a 1997 letter that the company would "focus relentlessly on our customers" and market leadership over "short-term profitability." It would make its employees sweat in the service of those customers. "It's not easy to work here...but we are working to build something important."[21]

Facebook and Google meanwhile offer most of their services—services billions of users rely on—for free. Users get secure email, the ability to create professional-looking documents, to stay in touch with family and friends, to build their brand, to express their artistic desires, to locate themselves nearly anywhere in the world and then tell everyone about it. And it costs nothing at the point of use. Isn't this what the internet dream was all about? It certainly doesn't sound like the Gilded Age titans who readily sacrificed the needs of the little guy for their own profit. Compared to them, the tech titans of today seem genuinely focused on what consumers need. They seem to have a big-picture vision of the future that's not focused on the bottom line.

Others insist that these companies' dominance won't last anyway. Michael Beckerman, president of the Internet Association, a trade group that represents Google, Amazon, Facebook, and Twitter, reassures us that "competition is just a click away."[22] Look at MySpace, or IBM, or Nokia, or AOL. No one is afraid of them anymore. By extension, given the speed of technological change, Facebook's, Google's, and Amazon's days of dominance must surely be numbered.

For all these reasons the new tech titans flew under the radar for a long time. To be sure, serious criticism surfaced in the years following the tech crash in 2000, but it didn't puncture the bubble of goodwill surrounding high-tech firms. No longer. Breach-of-privacy scandals, reports of psychological manipulation, and growing fears about the social implications of our love affair with smartphones have woken people up to the mammoths in our midst. "There's been a really big breakthrough," says Barry Lynn, director of the Open Markets Institute. "It's not just [coming from] the left. Interest in dealing with concentration of power, the fear of concentration of power is across the spectrum."[23]

Gathering Storm Clouds

The Democratic Party has traditionally been quite cozy with Silicon Valley—the unofficial capital of digital-technology development in the Santa Clara Valley south of San Francisco—particularly during campaign season. But leading Democrats, including Senator Elizabeth Warren of Massachusetts and former congressman Keith Ellison of Minnesota, have recently raised criticisms of America's new monopolies, with Ellison declaring, "They're too big."

Meanwhile, Bill Galston, a senior fellow from the Brookings Institution, and Bill Kristol, founder of the *Weekly Standard*, have begun a centrist policy project called the New Center that focuses heavily on technology. In a recent paper, they call for "comprehensive federal legislation" to regulate these companies, which "pervade our lives in ways that corporate behemoths

of years past never did."[24] For the first time in a long time, American monopolies and what we should do about them are being publicly debated.

The emerging criticisms of the tech titans are varied. Some denuncia-
tions seem to be simply the result of the Silicon sheen wearing off. With
fresh eyes it's clear that these companies, despite their innovative structures,
are doing many of the same things we criticized "old fashioned" companies
for, such as mistreating the people who work for them.

Apple was one of the first tech companies to receive this criticism, back
in 2010, when news came out that the Chinese workers assembling Apple
iPhones and iPads in subcontractors' factories were operating in a hellscape
of stress, drudgery, and exhaustion. Apple promised that it would address
the situation, but recent reports suggest that when orders are on the line
Apple is still willing to push the burden onto its least-privileged workers.
The *Financial Times* reported that student "interns" were being forced to
work extended overtime; the students attend a technical school that conditions graduation on putting in enough hours on the Apple assembly line.[25]

Readers may remember the 2015 *New York Times* exposé, "Inside Amazon: Wrestling Big Ideas in a Bruising Workplace," in which a former Amazon employee was quoted as saying, "Nearly every person I worked with, I
saw cry at their desk."[26] It wasn't the first time Amazon made the news for
making its workers cry. Spencer Soper wrote a series of investigative pieces
for the *Morning Call* in Eastern Pennsylvania detailing the difficult and precarious existence of Amazon's Allentown warehouse "pickers," who walk up
to twelve miles per shift in alternately sweltering or freezing warehouses.
During the summer, ambulances hang out in the parking lot ready for the
inevitable collapse. Many of these workers are "permatemps," workers who
engage in full-time, long-term employment at a company but don't receive
the benefits that permanent employees enjoy.

These days it's not uncommon to read stories about Amazon that detail
the human costs of our emerging smartphone economy. In the UK, Amazon
delivery drivers have to deliver up to two hundred packages a day. To accom-
plish this gargantuan task they average over eleven hours of driving per day.
If they don't meet their quota, they are fired, prompting drivers to desperate behavior, including speeding, and urinating and defecating in buckets in
their vans to avoid having to take a break. Amazon pays its delivery drivers
a daily rate, rather than an hourly wage, so despite their best efforts, many
drivers end up making less than minimum wage because it takes them so
long to complete the daily delivery quota.

Google has also been criticized for the conditions of its permatemps, who now outnumber Google's direct employees. During an active-shooter event at its YouTube headquarters in San Bruno, California in April 2018, only permanent employees received a warning text about the situation, prompting an outcry and broader scrutiny of Google's employment culture.[27] Moreover, while Google prides itself on its diverse and inclusive workplace, recent reports call into question its commitment to equality. Government vendors, of which Google is one, are randomly audited to make sure they are following federal diversity protocols. A random inspection by the Office of Federal Contractor Compliance Program found "systemic compensation disparities against women pretty much across the entire workforce."[28] The disparities, based on data from over twenty thousand Google employees, showed nearly seven standard deviations between males' and females' pay across job classifications.

Google workers were likely well aware of this state of affairs, but there is intense pressure inside the firm to present a sunny public face, as one email written by a Google exec demonstrates: "If you're considering sharing confidential information to a reporter—or to anyone externally—for the love of all that's Googley, please reconsider! Not only could it cost you your job, but it also betrays the values that make us a community."[29] Google workers are muffled by strict confidentiality agreements that likely violate federal and state employment law. In May 2016 the National Labor Relations Board filed a complaint against the company for "unlawful surveillance and interrogation in order to chill and restrict employee rights." Seven months later, the company's "spying program" to spot and eliminate leakers was the subject of a lawsuit, *Doe vs. Google, Inc. et al.*, filed in San Francisco Superior Court.[30]

Criticism is also growing about the impact of the tech titans and their business practices on the stakeholders in their communities. Facebook has been called out for flouting government regulations regarding discrimination in advertising. The quality and quantity of Facebook's data regarding the characteristics and preferences of ordinary people are unprecedented. Its platform is also cheap and easy to use; in minutes, a company or organization can place an ad on Facebook specifying exactly whom it wants (and whom it doesn't want) to see that ad. The problem is that advertisers can easily target groups in a way that flouts federal antidiscrimination laws.

For example, discrimination in the housing sector has long been a problem in the United States. Federal law prohibits real estate brokers and landlords from discriminating against potential buyers or renters on the basis of gender, sexual orientation, race, ethnic identity, or political beliefs. Yet the watchdog group ProPublica revealed that it was remarkably easy to place a housing ad on Facebook that does precisely these things. As an experiment ProPublica bought dozens of rental housing ads on Facebook, but limited which categories of users could actually see the ads: no African Americans, mothers of high school kids, people in wheelchairs, Jewish people, Spanish speakers, or expats from Argentina. They chose these categories because they represent protected categories under the federal Fair Housing Act. Ads that indicate "any preference, limitation, or discrimination based on race, color, religion, sex, handicap, familial status, or national origin" with respect to "the sale or rental of a dwelling" are illegal. Pretty simple. Yet, every one of ProPublica's ads was quickly approved. When the watchdog publicized how Facebook's ad policy flouted federal regulations the company was apologetic and promised to build an algorithm to fix the problem.

A year later, after Facebook had declared the problem fixed, ProPublica ran the same experiment and got exactly the same result. The ProPublica experiment might seem arcane, but as a major destination for ad spending, Facebook has come to play a key role in deciding who gets housing and who doesn't. In March 2019, the Department of Housing and Urban Development (HUD) sued Facebook for its ad policy, citing violations of the Fair Housing Act and discriminatory practices against users.[31]

Highly paid tech workers and the companies that employ them are also creating havoc in the places where these tech firms settle, such as East Palo Alto and San Francisco. In California, for each new tech worker, it is estimated that between two and five supporting jobs are created—dining hall servers, bus drivers, security guards, and cleaners. New jobs are great, but these jobs tend to be low-paid contract gigs. With the skyrocketing rents in communities near and in Silicon Valley caused by increased demand from highly paid tech workers, many ordinary folks find that they no longer can find a place to live in the region. Celia Cucalon, a housekeeper and eldercare worker, tells an increasingly typical story. Cucalon's landlord of nine years recently sold the building and handed Cucalon a notice stating that the rent for her 300-square-foot apartment would increase immediately from $1,035 a month to $2,250 and that she had 60 days to vacate if she didn't sign a new lease.[32]

Silicon Valley Rising, a local community-labor coalition fighting for people like Cucalon, says the quality of life in the region for the average family has deteriorated in recent years. Between 2000 and 2010, median family incomes fell 20 percent. While well-paid tech workers get gleaming white buses to shuttle them to their jobs and home again and all-inclusive campuses with bowling allies, free food, dry-cleaning, and hair salons, working-class Valley residents get unaffordable housing, deteriorating schools, and mushrooming homeless encampments.[33]

These companies aren't solely responsible for painful swings in the real estate market, but they do have a significant impact, which they have not acknowledged in a meaningful way. Facebook, Google, Apple, and the dozens of other tech companies operating in the Valley are extremely profitable. Tech companies could plow some of these profits back into the communities they are "disrupting" or use their clout to help develop serious solutions to the region's housing crisis. But the overwhelming trend is not to put profits or influence to work for ordinary people. Instead, these tech titans use a variety of accounting tricks—such as creating "sock-puppet subsidiaries" to shift profits to low-tax locations outside the United States—to avoid paying their fair share of taxes.[34]

The issue of corporate tax evasion is not a new problem. Since the mid-1990s, multinationals based in the United States have increasingly shifted profits to offshore tax havens. Gabriel Zucman, an economics professor at UC Berkeley, says American multinationals hide 63 percent of the profits they claim to earn overseas in a handful of tax havens such as Bermuda, Luxembourg, and the Cayman Islands.[35]

But Silicon Valley firms are masterful avoiders of the tax man. Facebook made over a billion dollars in profits in 2012, the year of its IPO (initial public offering), yet used an accounting trick to avoid paying even one cent in US taxes. Bezos designed his company with tax avoidance in mind, initially planning to open Amazon on a Native American reservation in California. That plan was scrapped, but he developed a better one. Because of a loophole in the federal tax code, Amazon didn't have to pay state taxes in the states to where it shipped its books (and later, millions of other items) as long as it didn't have a brick and mortar store in those states. Estimates put Bezos's tax savings from 1994 to 2015 at over $20 billion. The loophole has been closed, but Amazon still managed to pay no federal tax in 2017 or 2018 on billions in profits.[36] Apple was handed a huge tax bill from the European Union, charging the Cupertino-based company with underpaying taxes in Ireland to the tune of $15.4 billion. It eventually paid the bill, but not before moving to a new corporate home—the Channel Island of Jersey, where corporate income isn't taxed. As of 2015, Google had "permanently reinvested" nearly $60 billion in profits in foreign tax havens, paying nothing on these profits to the US government.

Tax avoidance, discrimination, and oppressive working conditions are features that call into question the growing power of the tech titans to shape life in our emergent smartphone society. But these fairly straightforward misdeeds are just a piece of the problem. The rise of companies such as Facebook and Google has also created and exacerbated a host of more complex problems that we're just beginning to grapple with.

Thornier Problems

"Fake news" is a problem that's been around at least since newspaper publishers, Joseph Pulitzer and William Randolph Hearst, went head to head in the 1890s, spinning fantastic tales to see who could sell the most newspapers. Today, fake news—stories that try to mislead the reader with hoaxes and deliberate misinformation in an effort to discredit someone or for financial or political gain—are a topic of conversation once again.

On the surface, the problem of fake news is fairly simple. Increasingly, people get their news from Facebook, Twitter, YouTube, or Google News. Sixty-six percent of Americans get "at least some" of their news from social media; 45 percent go to Facebook for news, and for people under the age of fifty, the percentage is much higher.[37] Instead of picking up a newspaper from the corner store, many of us pull out our phones and scroll through our newsfeed. Woven in among posts from friends, rants from acquaintances, and fluffy pet pics we see all kinds of news stories: some are from familiar, and reputable, sites such as the *New York Times* and the *Wall Street Journal*, while others are from who-knows-where and, upon closer inspection, are fake. These apocryphal tales appear alongside real news stories in our feeds and are written like real news stories so people have a hard time telling them apart. Fake stories often make salacious or shocking claims, so readers share them, and before long they've gone viral.

The impact of these stories varies. During the 2016 presidential campaign, voters read and shared a bunch of weird, untrue stories—Hillary Clinton's pedophile ring in a pizza shop, Pope Francis's endorsement of Trump—that some experts believe swayed voters enough to give Trump the win. Other times the results are much more grim. Facebook garnered millions of new users in Myanmar through its Free Basics program, and as a result became a primary source of news in the country. Unfortunately its platform has also been used by the country's military as a tool for ethnic cleansing. Military officials set up fake celebrity pages and then used these pages to spread lies against the country's Rohingya Muslim minority. The psychological warfare sparked such violence against the Rohingya that 700,000 fled the country in 2017.[38]

Many have demanded that Facebook eliminate the fake news popping up on its website. Yet even though millions rely on Facebook for news, the company claims that it isn't a media company akin to a trusted newspaper. It says it is merely a platform for individuals, groups, and businesses to share content, and that beyond certain guidelines, it isn't interested in policing speech on its website. This response isn't surprising. If Facebook did admit it was a media company, it would have to act like one, vetting stories, checking facts, considering ethics—all the boring stuff that a company like the *Washington Post* or CNN does. By not self-identifying as a media company Facebook avoids liability for any nastiness that pops up on its site: section 230 of the Communications and Decency Act enables companies such as Facebook to avoid "intermediary liability" for the things people say or do on their platforms and, on the flip side, allows them to act as "Good Samaritans," policing their sites as they see fit.

Nonetheless, after a year of denying that fake news was a problem, Facebook quietly started reducing the amount of news people see in their newsfeed. Does this mean we can just go back to reading news wherever we read it before? Unfortunately not. Fake news is a manifestation of a much bigger problem—the replacement of quality, investigative journalism with "content."[39] This trend has radically accelerated with our move to the small screen.

As the digital has colonized our lives, and the tech titans have colonized the web, our digital lives have been hemmed in to a handful of apps. We leave these apps less and less because they're where our networks are and because they give us everything we need: messaging, photos, video, and news. Where our eyeballs go, advertisers go, creating a virtuous circle for the creators of these apps. The tech titans facilitate both the creation and sharing of content, using algorithms to show us what they think we'd like to see; and the more we use their platforms, the better their algorithms get at guessing.

For these titans, engagement is the priority, keeping us on their apps with listicles, celebrity gossip, and sometimes, real news. It doesn't really matter what the content is as long as we're engaged. Hard-hitting journalism is nice but it takes a backseat to clicks. With genuine news organizations allocating less and less money to paying professional journalists and editors, the "news" we read instead is a few good stories, usually posted over and over, mixed with an endless stream of algorithmically organized crap.[40]

A thriving press has long been considered crucial to a properly functioning democracy, providing, Jürgen Habermas writes, "an independent realm, a public sphere, a commons where citizens could meet to discuss and debate politics as equals."[41] This sphere is disappearing, replaced by an online space controlled by a few megacorporations. Emily Bell, the director of the Tow Center for Digital Journalism at Columbia University, contends, "Our news ecosystem has changed more dramatically in the past five years than perhaps at any time in the past 500." A few companies now control the news many of us see, and we're not even sure about the nature of this control because it is "filtered through algorithms and platforms which are opaque and unpredictable. . . . There is a far greater concentration of power in this respect than there has ever been in the past."[42]

A related problem concerns the nature of online search. It's a problem often associated with Google because Google *is* search. The story of Page and Brin's extraordinarily clever invention is well known. In the mid-nineties, as graduate students at Stanford, they created an algorithm, called PageRank, that ranks webpages by how many other pages are linked to it and how many people have visited that page. It was fair and useful. Suddenly you could find stuff that you wanted to find on the internet. We all started using Google. Google became a verb, "to google." We used Google to find everything, and soon everything could be found on Google.

Early on, Google investors demanded that Brin and Page add advertising so that their search engine would be not only useful but also profitable. At first Page and Brin demurred, saying that advertising would corrupt their mission. But in a for-profit economy, and an internet environment where nominally free services are the norm, this adless model soon came under pressure. Google created AdWords and then AdSense to sell ads, and began to give preference to companies and webpages who paid to be located at the top of the search results because most people never look past the first page of results.[43] Results from Google searches that appear neutral but have actually been prioritized because they were paid for slowly became the norm.

The European Union's eight-year antitrust investigation into Google, led by Margrethe Vestager, Europe's competition commissioner, gives a sense of just some of the problems associated with Google's monopoly. In one case, the Google Shopping Case, the EU Commission fined Google

€2.42 billion on an antitrust charge for abusing its near monopoly in online search to "give illegal advantage to its own shopping service," "deny[ing] other companies the chance to compete" and leaving customers without a "genuine choice." A second investigation examined whether Google unfairly bans competitors from websites that used its search bar and ads. The third case focused on Google's Android software for mobile phones, exploring how Google forces mobile phone providers and users to use its Android software and app store. In July 2018 the commission found that Google had abused the dominant market position of its Android operating system, fining the company a record €4.3 billion.[44]

These cases highlight how reliant we are on Google. Google controls 90 percent of search in both the United States and the European Union, while 60 percent of Americans use Google's Chrome browser. The cases also demonstrate how little we know about Google's ranking system and algorithms. Google's design and ethos encourages us to see that peaceful search bar as a neutral arbiter of web content. But investigations like the ones led by Vestager call this feeling into question.

Another wrinkle in this story is personalization—what Eli Pariser, now the chief executive of Upworthy, defines as the "invisible algorithmic editing of the web." When we pull out our phones to search for, say, the presidential election, we most likely get very different results from another person sitting next to us doing an identical search on their phone. The information we receive is tailored to us on the basis of our past searches, websites visited, our current location, what type of phone we're using, and a host of other factors. As Pariser says, we each live in a "filter bubble"—our "own personal unique universe of information." The idea of a standard Google in which everyone typing the same query gets the same search results simply no longer exists.[45]

Personalization is not limited to Google; Facebook, Yahoo, Twitter, even the websites of major newspapers are using personalization to increase user engagement. OK, so what? Isn't it great to see what one wants to see? Perhaps. But it's also worth being exposed to things that make us uncomfortable, worth knowing what's happening in the world around us. More important, we should have a clear understanding of what we're getting when we use a tool such as Google Search. But we don't. We have no idea how our "personal unique universe of information" is constructed, what's let in and what's kept out. We're stuck with new gatekeepers—the tech titans.

Our phones link us 24/7 to a digital world of ideas and information. We've come to rely on Facebook, Google, and, increasingly, Amazon as we would rely on the lights turning on when we hit the switch, or the water coming out clean when we turn on a faucet. Facebook and Google have created the equivalent of utilities—infrastructure that we depend on, that the economy as a whole depends on.[46] Our social, political, and economic life as we know it today would not be possible without the software and algorithms created by Google and Facebook.

We may treat the architecture these tech titans have created as a digital commons, but it is not a commons. These companies are privately owned and run for profit. A few tech firms have outsized, and unprecedented, control over our economic, social, and political life, yet we have very little control over them. Despite the importance of this infrastructure to our way of life, we have largely left these companies to govern themselves. Like the titans of the Gilded Age, they operate in an extremely lax regulatory climate. We have put almost no legislation in place to demand accountability, we have no checks and balances, few mechanisms for examining how these companies are reforming themselves in response to criticism, problems that arise, or their quest to increase profits and market share. We seem to be fine with vague pronouncements and promises to fix problems as they arise. We don't follow through to see if problems have actually been resolved. Instead we rely on a few dogged tech journalists, experts, and think tanks to troubleshoot.

Rebooting the Gilded Age

We've been here before. The history of the Gilded Age captains of industry's using their control over finance and industrial infrastructure to line their pockets and gain unprecedented power over farmers, factory workers, and politicians is well known. Indeed, these days Zuckerberg, Bezos, Brin, and Page are often compared to Rockefeller, Morgan, Carnegie, and Vanderbilt. So how did we let ourselves get into this position again?

In part, these companies have benefited from what we might call an ideologically favorable climate: neoliberalism. They have grown up in an environment of ideas and policies in which government regulation and stakeholder oversight are viewed as suspect, one in which "efficient" markets are assumed to encourage effective corporate self-regulation.[47]

The tech titans are also very powerful, especially where they are located, making cities and communities afraid to challenge them for fear of losing investment and jobs. When Seattle tried to implement a head tax to fund affordable housing, Amazon, which controls nearly 20 percent of Seattle's office space, pitched a fit and the city backed down.[48] (Amazon's recent exit from Queens after local opposition to the company's sweetheart tax deal could signal a shift in tech companies' power to steamroll local constituencies, however.[49])

Equally important is the money and effort spent by these companies to mold public opinion. A 2017 *Wall Street Journal* investigative report found that "over the past decade, Google has helped finance hundreds of research papers to defend against regulatory challenges to its market dominance."[50] These papers, written by academics and think tanks, have a real impact: they're distributed to policy makers, staffers at regulatory agencies, and congressional committees.

Google provides much needed funding to third-party organizations— the American Library Association, the American Association of People with Disabilities, the National Hispanic Media Coalition, and the Center for American Progress, to name a few.[51] It also funds academic institutions and media fellowships, and even organizations such as the Electronic Frontier Foundation and the Computer and Communications Industry Association, which provide expert analysis and purportedly unbiased opinion on tech-related matters for the general public.[52]

Google also spends heavily on lobbying. The Benedictine Sisters of Baltimore is one of several groups of Google shareholders who recently submitted a proposal to the company demanding to know how much money it spends on lobbying. The answer: Between 2010 and 2015, Alphabet, Google's parent company, spent $80 million on lobbying just federal lawmakers. That doesn't count money spent lobbying individual state lawmakers, trade associations, and overseas governments. Google also leases a 55,000-square-foot office less than a mile from the Capitol, no doubt facilitating the 128 visits to the White House made by Google executives during the Obama years—more than any other corporation.[53]

Word on the street is that former Google CEO Eric Schmidt learned a valuable lesson from Microsoft's antitrust case: Microsoft hadn't spent the time and money generating goodwill in DC and it paid the price. No one could accuse Google of the same, certainly not after events like its top-secret Google "summer camp," held at a luxurious Sicilian resort. The yearly event is designed for world business leaders and tech gurus to discuss global problems, policy proposals, and the future of the internet. Guests enjoy culinary tours, golf, Thai massage treatments, and the company of celebrities such as Emma Watson, Sean Penn, Prince Harry, and Sir Elton John.[54]

Google cultivates more than goodwill with its generous funding. It also keeps opinion makers in line. This was recently made apparent when New America, a Google-funded think tank in Washington, fired Barry Lynn and his entire team of researchers two days after Google threatened to pull its funding from the center, allegedly in retaliation for Lynn's public support for a recent EU ruling against Google. Lynn believes he was fired for speaking out.[55]

Google is not the only tech company greasing the wheels in Washington. The Center for Responsive Politics reports that in 2016, the internet and electronics industry together spent a record $178.5 million on federal lobbying, putting them right behind the pharmaceutical industry and twice Wall Street financial firms' spending.[56] Continuing the revolving-door tradition between Washington and the internet industry, Facebook hired for high-ranking positions Chris Herndon, former counsel on the Senate Commerce, Science, and Transportation Committee; Caitlin O'Neill, Nancy Pelosi's former chief of staff; and Jodi Seth, John Kerry's communications director.

Tech titans are also testing other strategies to ward off growing criticism. Mark Zuckerberg went on a "listening tour" of the United States beginning in January 2017. In the months after the presidential election, as public disapproval grew over the company's role in allowing fake news to possibly impact the election results, Zuckerberg declared that he needed to get out to the people. He posted highlights of his adventure on his Facebook page: he fed a calf, went to Mardi Gras, met with church congregations, did photo ops with basketball players from the Duke University and University of North Carolina teams, had a sit-down with a group of recovering heroin addicts, pet a wildlife service dog in Glacier National Park, went to a Texas rodeo, and met with the Blackfeet Indian Reservation's tribal council.

Zuckerberg also listens to his tech predecessors. Unlike the negative lesson Google learned from Microsoft, Zuckerberg learned a positive lesson from Bill Gates: the power of technophilanthropy. Along with his wife, Priscilla Chan, Zuckerberg established the Chan Zuckerberg Initiative (CZI), backed by a promise to give away 99 percent of the couple's net worth in their lifetimes. CZI has funded a variety of causes, including college tuition for undocumented students, and Chan's pet project, a private, tuition-free elementary school in East Palo Alto.

In this way, the new titans are following in the footsteps of the old titans, many of whom became heavily involved in philanthropy after building their fortunes. According to the historian Jackson Lears, in his later years,

John D. Rockefeller "embraced a doctrine of stewardship," believing "God had entrusted him with his money for disbursal in accordance with the divine will." Andrew Carnegie developed his own secular philosophy of philanthropy—his "gospel of wealth" which "insisted that it was a disgrace for a man to die rich and that only great munificence could justify great wealth."[57]

Those old titans may have been assuaging the guilt they felt for shunting aside people and morals in their quest for wealth and power. But they were keenly aware, and fearful, of mounting criticism from the Populist movement and their increasingly restive workforces whose actions were gaining widespread sympathy from the public. Rural Populists and urban Progressives, through trial and error, success and failure, formed an antimonopoly movement that fought against the unchecked power of America's titans and also *for* a new vision that demanded new institutions and legal regimes to protect consumers and small producers.

Are we on the cusp of a similar shift today? It's hard to say. To get a sense of the future, we must first examine how smartphones are not only making a new class of titans rich and powerful beyond anyone's wildest dreams, but are also at the center of a new frontier of capitalism.

Chapter 3

New Frontier

"Nude selfies till I die."

Kim Kardashian's Webby Award acceptance speech for "excellence on the internet" was totally #goals. It also speaks to the reconfiguration of the public and private in our smartphone society. Kim and her selfie-obsessed sisters post sultry snaps of themselves online—taken in their bathrooms, bedrooms, and cars, on family vacations and shopping trips—and millions of people devour them. The formula is robust. *Forbes* named Kylie Kardashian the "world's youngest self-made billionaire" at twenty-one.

The demand for all things Kardashian is so vast that Kim published *Selfish*, a 500-page book consisting solely of selfies: Kim at a red light; Kim at her fragrance shoot; Kim at the Yeezus show. *Selfish*, like all Kardashian content, triggered "feelings." Laura Bennett from *Slate* "can't recommend it enough"; despite Kim's "aggressively repetitive" captions, Bennett found the book "riveting."[1] Others were less impressed. *New York Daily News* staffers staged a mock dramatic reading on YouTube. But the gag fell flat. Perhaps no amount of irony can make "Another bikini Thailand selfie" funny. Or perhaps we suspect the joke is on us: the Kardashians have cashed in on what most of us give away for free.[2]

Actually, it was Kim's "momager," Kris Jenner, who figured out how to "monetize that shit!" as digital marketing guru Gary Vaynerchuk would say.[3] Jenner created an empire from a few connections and her daughter's sex appeal that includes a hit reality show (*Keeping Up with the Kardashians*), several profitable beauty and clothing lines, and smartphone apps to follow each sister's life.

The Kardashian clan is not alone in its social media savvy. Like the fairytale stories of small-town kids who found fame and fortune in Tinseltown,

YouTube, Twitter, and Instagram have turned Daniel Middleton, Kelly Oxford, Huda Kattan, and hundreds of others into microcelebrities. One anonymous YouTuber—you can only see her hands and hear her voice—makes millions of dollars a year from videos that show her hands unwrapping, assembling, and playing with Disney toys.

Of course, relatively few of us are cultivating our social media biopics for the Benjamins. Yet, if we're honest, many of us rival the Kardashians in how "extremely online" our lives have become. Three billion people a month spend an average of 135 minutes a day on social media—Facebook, Instagram, YouTube, Snapchat, Pinterest, WhatsApp, Twitter, to name a few—and 70 percent of our social media time is spent on our phones (total screen time stretches considerably longer). Social media experts say that "to decouple social media from mobile use is impossible."[4]

Much social media content is thematically similar to Kardashian fare. Instagram and Twitter are bottomless receptacles for our lovingly crafted "squinty" and duck-faced selfies snapped in restaurants, parks, museums, funerals, ambulances, and concentration camps. We're so enamored with our mugs that we'll risk meeting our demise for the perfect snap; a spate of selfie-induced accidental deaths has led cities to post warning signs urging visitors to put away their selfie sticks near cliffs and water. In India, Mumbai and Goa have designated "no selfie zones" in popular tourist spots because people keep falling and drowning while posing for their own cameras.

Social media is more than smiling faces, however. It's as if humanity had been waiting its entire existence to post mansplaining DIY videos and makeup tutorials, to pin dream kitchens and rainbow cupcakes, to write pedantic reviews and vicious tweets, to share cat hilarity, celebrity memes, and letters telling off bridesmaids who wouldn't shell out for a destination wedding. The amount of time and energy we spend posting, snapping, creepin', sharing, trolling, and scouring is mind blogging. It gives new meaning to the truism that humans are social creatures.

Over the past decade, as smartphones have become ubiquitous, a shift has occurred in the boundary between the public and private. Our hand machines have enabled us to blend the digital and analog elements of life into new configurations, creating new social norms—the patterns and rules of interaction.

On the most basic level we are prepared to share, and expect others to share, personal information online that we never would have ten or fifteen years ago. Chris Rojek, a sociology professor at the University of London, observes that "private life is now, more than ever before in the modern world, lived in public."[5] The legions of Queen Bey fans reflecting on Beyoncé's latest Instagram pic demonstrate how new rituals—"mutually focused activities that engender a common mood in a bounded group"—have emerged that occur only online.[6] Meanwhile, old rituals are being reshaped. Instead of the yearly Christmas card update we maintain a constant stream of family pictures and news.

More than this, we imagine ourselves differently. We see ourselves as members of digital networks outside our communities and families. These networks feel real. For billions of people, participation in them has meaning and impact in shaping their identities. Social life itself becomes interpreted through these networks as new interactions and wanderings are woven with both analog and digital threads. The more we're online, the more life is designed to be online.

All this socializing raises a somewhat mundane question: Relatively few of us share the mercenary aims of the Kardashians, so why *are* we spending so much time posting perfect pics of our fur baby, wondering why the one we just posted of the little guy with birthday cake on his nose isn't getting noticed, scouring Instagram for other dog lovers, and watching YouTube compilations of canine hilarity?

It's complicated.

Consider Instagram. About 33 percent of Americans (and about 70 percent of people under thirty) say they use the Facebook-owned photo-sharing site. But even though US users account for the largest single share, 80 percent of Instagram users live outside the United States, making it a truly global app. Nearly half of Swedes, 40 percent of Turks, and roughly 20 percent of Japanese people use the app.

Using the site is straightforward, but summing up what people do with it is less obvious. Mori, a thirty-two-year-old Japanese furniture company employee, is an avid user. Recently Mori has been really into pancakes, particularly pancakes with maximum *fuwa-fuwa*, fluffiness. So far Mori and her friends have tried over six hundred pancake houses in Japan. Each trip includes the same ritual—taking pics while ordering the pancake (seasonal is best), taking pics while pouring the syrup, taking pics while testing for fluffiness, and then posting the pics to Instagram.[7]

Mori's Instagram ritual is not unusual. Japanese businesses offer a growing array of experiences designed with *insta-bae* principles in mind—a portmanteau word describing something (a plate of food, an outfit, a beautiful setting) that you just know would look amazing on Instagram. One café blends coffee and birds. Enthusiasts reserve a spot months in advance to snap a picture of themselves drinking a latte while petting a snowy white owl chained to their table.[8] Documentation has become central to the "experience economy" in which the memory or experience of the event is the product being sold.[9] Did you pet the owl if there's no picture of you doing it?

Restaurants and bars in American cities have also learned to sacrifice to the Instagram gods. Patrons of Nobu Malibu, an upscale sushi bar run by Nobu Matsuhisa (an "Iron Chef" from the eponymous Japanese cooking show) in Malibu, California, post hundreds of nearly identical snaps of artful plates of sushi and beautiful sunsets off the deck. The restaurant 230 Fifth, the most Instagrammed eatery in the country, shows up in endless pictures of people sharing drinks in the rooftop winter dome. CatchLA has a beautiful indoor garden prompting a wide selection of posing-under-the-wisteria shots.

These pictures show both the wide appeal of #eatstagram and also how our experiences are simultaneously performances. Instagram enables the status-conscious to show they hang out in cool places—to demonstrate their cultural capital, or habitus, as the French sociologist Pierre Bourdieu might say.[10] Comedian Tim Dillon jokes that you can tell someone's social class by the summer pictures they post on the site: rich people post pics of the Hamptons and pool parties; middle-class people show their bar crawls, babies, and Tough Mudder races; poor people post memes and snaps of French fries.

Humans have long been obsessed with managing the impressions they make on others. The Nobel Prize–winning author Alice Munro articulates this managerial impulse perfectly in her 1973 short story "Material." Browsing through a bookstore a woman comes upon a new volume of stories written by her ex-husband, Hugo. Examining his picture and the biographical blurb below it, she ponders the "shop-worn and simple" disguises, or identities, that people take up. She laughs at the "lies, the half-lies, the absurdities" that populate Hugo's bio—that he "has worked as a lumberjack, beer-slinger, counterman, telephone lineman, and sawmill foreman." Having once been married to the man, she knows for a fact that he was never a telephone lineman. "He had a job painting telephone poles. He quit that job in the middle of the second week because the heat and the climbing made him sick." She wryly wonders why he chose to exclude "examination marker" from his bio—the job he got after quitting the pole-painting job and one that he liked much better. Nor was he ever a sawmill foreman. He worked at his uncle's mill one summer; "what he did all day was load lumber and get sworn at by the real foreman."[11]

Munro's story still resonates. Social media is full of "lies, half-lies, and absurdities"; it is the place where we perform our biographies. In his seminal work *The Presentation of Self in Everyday Life*, the sociologist Erving Goffman argued that social interactions can be thought of as performances, and that people's performances vary depending on their audience. We enact "front-stage" performances for people—acquaintances, coworkers, judgmental relatives—we want to impress. Front-stage performances give the appearance that our actions "maintain and embody certain standards." They convince the audience that we really are who we say we are: a responsible, intelligent, moral human being. But front-stage performances can be shaky and are often undermined by mistakes—people put their foot in their mouth, they misread social cues, they have a piece of spinach lodged in their teeth, or they get caught in a lie. Goffman was fascinated by how hard we work to perfect and maintain our front-stage performances and how often we fail to do so.[12]

Smartphones are a godsend for the dramaturgical aspects of life. They enable us to manage the impressions we make on others with control-freak precision. Instead of talking to each other, we can send text messages, planning our witticisms and avoidance strategies in advance.[13] We craft a Tinder persona that says "earnest yet sexy." Our Facebook profile projects A Serious Thinker, unconcerned with the gaining status among colleagues.

There are more layers to our social media obsession, however, than the desire for experiences, status, and social control. People also seek connection, meaning, and community. We watch Susan Boyle sing "I Dreamed a Dream" in her 2009 *Britain's Got Talent* audition for the eighth time for more wholesome reasons. Humans desire authenticity; surprise and the unexpected triumph by someone who seems to be like us trigger a deeply satisfying feeling—a surge we crave and attempt to recreate.

Social media offers a sense of connection, even if it's with people rarely or never met. For isolated individuals, such as elderly Americans, this sense of connection is nontrivial. In contrast to assumptions that the elderly are slow on the uptake when it comes to technology, social media use (particularly Facebook) has grown rapidly among older generations in the past few years. Elderly Americans use social media as a way to fight loneliness and as an important source of social and emotional support; it helps them reach out to or stay in touch with family and friends, and more broadly to stay engaged with life—a key factor in "successful aging."[14]

Networks such as the "mamasphere" offer a different kind of support. The mamasphere is comprised of thousands of pregnancy apps, parenting blogs, and humorous Instagram pages about the trials of motherhood. @the realramblinma offers irreverent nuggets such as "It's not nighttime until you tell your kids to 'GO BACK TO BED' in your best death metal voice." The popular photo-sharing site Pinterest reaches 83 percent of American women between the ages of twenty-five and fifty-four, and is the holy grail of perfect mothering—a place you can gorge on images of decluttered playrooms, fun lunch ideas, and birthday party theme ideas for ten year olds.

The mamasphere is quite performative: like we're ever going to make an artichoke pesto with parmesan and salami whole grain roll-up with fresh blueberries and roasted chickpeas for our kid's lunch. No child actually enjoys playing with vegetable-dyed wooden toys, and that sparkly silly putty is not coming off the couch. Despite its theatricality, however, the mamasphere is also a space of solidarity. Communications arts professors Julie Wilson and Emily Chivers Yochim contend that "family life feels increasingly precarious and mothers are responsible for managing threats to their families' economic, emotional, and physical well-being." This is a daunting responsibility. Many of us have days where we really doubt ourselves as parents. The mamasphere can offer support and reassurance that others are struggling too. Wilson and Yochim say it creates communities and provides "a space for sharing and socializing" that transforms the "privatized domain of domesticity into an ongoing mediated collective endeavor."[15]

This transformation of the private sphere into an "ongoing mediated collective endeavor" is also apparent in how some members of the LGBTQ community, particularly teens, use social media. While distressing stories of vulnerable teens being cyberbullied abound, social media also offers a safe space, particularly when accessed through a private smartphone, for LGBTQ teens to not only connect with others in the community, but also to engage in a controlled exploration of their sexuality and gender. They can choose to be out online but not at home, to adopt a different identity online, or to use an online platform to inform their networks after they've decided to come out at home.[16]

In addition to community our digital interactions provide us a sense of meaning. Rojek's work on the rise of celebrity culture is useful here. "Celebrities offer peculiarly powerful affirmations of belonging, recognition, and meaning in the midst of the lives of their audiences, lives that may otherwise be poignantly experienced as under-performing, anti-climactic or sub-clinically depressing." Para-social interactions with celebrities, Rojek writes, "enjoin us to adjust to our material circumstances and forget that life has no meaning."[17] Katie Heaney, a senior writer for *The Cut*, opened up about her secret obsession with creepin' on strangers on Instagram: "I follow them not because I'm a genuine fan, but because their social media presence provides me with the dramatic twists and turns my own life generally lacks."

Social Media Is Bad

Social media may bring a sense of meaning and community, but at what cost? From psychologists to tech experts, a growing chorus of voices calls for a return to "real life." What seemed like a fun, positive thing is actually terrible, we're told, turning us into narcissistic weirdoes and possibly destroying society.

We worry that social media is changing our subjectivity—that we've become so attuned to "likes," retweets, and follows that our self-esteem begins to depend on them. If a picture doesn't get enough "likes," we disappear it. If our tweet doesn't get retweeted, we feel a bit deflated. Feelings of inadequacy begin clawing at our leg. We start using that selfie software to look thinner, more luminous. We spend hours thinking of clever memes. Our conception of our self becomes inseparable from the social media story we construct.

In the midst of this intertwining many of us become addicted to our small screens and the worlds they open up. Granted, a recent Pew Charitable Trust report found that more than half of adult social media users believed they could "quit easily," but the number of people who say it would be "difficult" to quit has increased.[18] Popular accounts certainly emphasize the challenge of putting down our hand machines. Journalist Andrew Sullivan wrote a moving piece for the *New Yorker* about his own addiction, which got so bad that he eventually checked himself into a ten-day digital detox session at a former novitiate in Massachusetts in an effort to regain a sense of equilibrium. Handing over his phone wasn't easy: "I duly surrendered my little device, only to feel a sudden pang of panic on my way back to my seat. If it hadn't been for everyone staring at me, I might have turned around immediately and asked for it back."[19]

Sullivan's always-connected life, much of it tied to his work as a popular blogger and writer, had become unsustainable. His health and emotional well-being seemed to deteriorate in direct proportion to the amount of time he spent tweeting, blogging, reading, following. His smartphone and the digital world it connected him to had him jumping like "a witless minnow," slowly banishing all those "spaces where we can gain a footing in our minds and souls that is not captive to constant pressures or desires or duties." There was no time for reflection, no time to quiet the mind, and it was killing him.[20]

Sullivan is not alone. Bernard Harcourt, a law professor and critical theorist, says the digital stimulation brought by our smartphone apps "[taps] into our lateral hypothalamus, triggering the kind of addictive 'seeking' that absorbs us entirely, annihilates time, and sends us into a frenzied search for more digital satisfaction."[21] The growing consensus is that we've lost control—that we're fast becoming like those dead-eyed slot-bots with their quarter cups and fanny packs slumped over the machines at an off-strip casino.

If we know this is happening—after all, knowledgeable people keep warning us—why don't we just stop? Graham Dugoni, founder of Yondr, a manufacturer of individual neoprene cases that entertainment venues use to lock up people's smartphones, says, "Our attachment to our phones isn't all that intellectual. It's much more a body thing."[22] Nir Eyal, the author of *Hooked: How to Build Habit Forming Products*, agrees: "Feelings of boredom, loneliness, frustration, confusion, and indecisiveness often instigate a slight pain or irritation and prompt an almost instantaneous and often mindless action to quell the negative sensation." We can't help ourselves. We reach for the phone.

More and more, people are at war with their phones. Heaney—the one who enjoys creepin' on Instagram—talks about skirmishes between her "rational brain" and her "emotional brain": "My rational brain finds the social media endless-joy-gusher infuriating. My emotional brain can't get enough." Freelance writer Luke O'Neil seems to have lost the smartphone war; O'Neil says he suffers from Internet Broken Brain as a result of his extremely online life and likens himself to a "scumbag" drug addict.[23]

Recently, members of the tech community have stepped up to shoulder some of the blame for these problems. Confessing their sins, software engineers describe how the platforms they helped create exploit basic psychological tricks to keep users hooked. Leah Pearlman, one of the people who created Facebook's "like" button, says she feels uncomfortable with what she has created and admits her own susceptibility to its "addictive feed back loop."[24]

Pearlman is not the only former Facebook employee who's come out against the social media company. Chamath Palihapitiya, a former Facebook VP who now runs a "mindful" venture capital fund, doesn't pull any punches. In an appearance at the Stanford Graduate School of Business he told the audience, "The short-term dopamine driven feedback loops we've created are destroying how society works."[25] Facebook's first president, Sean Parker, brags about how the company uses "little dopamine hits" to "consume as much of your time and conscious attention as possible."

Former Googler Tristan Harris, one of a growing chorus of tech refuseniks, channels sci-fi pioneer William Gibson when he says, "All of us are jacked into this system. All of our minds can be hijacked. Our choices are not as free as we think they are."[26] Guillaume Chaslot, also a former Google employee, says YouTube's recommender algorithm keeps people glued to the page by showing them increasingly extreme content.[27] Chaslot claims that he was fired for pushing too hard to get the company to address his ethical concerns regarding the algorithm.[28] Jaron Lanier, a longtime tech insider, recently published a book, *Ten Arguments for Deleting Your Social Media Accounts Right Now*, in which he denounces social media and warns readers that they are destroying our character, our ability to empathize, our soul.[29]

The growing consensus is that we're in thrall to a technology that we are psychologically incapable of handling. Our failure to control our impulses has made us less human. Sullivan compares himself to a fish and titled his confession, "I Used to Be a Human Being." Columbia Law School professor Tim Wu likens us to B. F. Skinner's pigeons, while Lanier proclaims, "We're all lab animals now." Siva Vaidhyanathan, a media scholar and vocal critic of Google, implores us to "rehumanize ourselves."[30]

The fear of becoming less human and more beastly because we can't control our impulses is rooted in an old set of ideas commonly associated with René Descartes, a seventeenth-century French philosopher. Descartes viewed the body as a beast that must be constantly reined in and controlled, the way rulers govern unruly subjects; the senses are seen as a "prison for the reasoning soul." In this framework, our "true" self—our soul or spirit—is something apart from and more pure than the body.[31]

Losing control, and with it authenticity, has often been linked with technology. Adam Smith feared machines would turn us into monstrosities. Karl Marx spoke of humans becoming appendages to machines. In the waning years of the nineteenth century, intellectuals were distressed by the rise of cinema, a form of entertainment that "operat[ed] on the emotions and viscera, on the seats of irrationality and irresponsibility," says the media historian Neal Gabler. Critics decried cinema as "sensational" (a pejorative back then), "beyond the reach of intellect," and a dangerous pastime that "excites the nervous system in much the same way drugs did."[32]

The mind-body divide was front and center in these critiques. Popular entertainment was said to generate Pavlovian responses; its consumption was unthinking, addictive. The old cultural order was being overwritten by a new, inferior one; the sublime was being replaced with fun.[33] These framings are echoed in how we talk about smartphones and our use of social media today: once again, a new machine has turned us into unthinking automatons, driven by desire, impulse, and algorithms rather than reason and thought.

But we should be careful about how we frame our fears. It is easy to fall into mythmaking. Donna Haraway, a prominent technology scholar and feminist, has long warned about the limitations of imagining a pure human who existed "before" technology, the misconception that our hunter-gatherer selves, or our agrarian selves, were somehow more human than our smartphone selves.[34] Sociologist Eva Illouz notes how much of our critique reflects a "longing for purity," and certainly the way debates about smartphones are framed reflects this obsession with purity and our "true selves."[35] The pervasive distinction between "real life" and "digital life" reflects this most of all: the idea that one is real and the other is simply performance or fantasy. But there is no pure life apart from our interactions, whether they be digital or analog. As Erving Goffman said long ago:

> To the degree that a performance highlights the common values of the society in which it occurs, we may look upon it . . . as a ceremony—as an expressive rejuvenation and reaffirmation of the moral values of the community. . . . To stay in one's room away from the place where the party is given, or away from where the practitioner attends his client, is to stay away from where reality is being performed. The world, in truth, is a wedding.[36]

The point here is not to dismiss our fears and critiques of social media and smartphones. On the contrary, we must take these fears very seriously and not allow them to be pigeonholed into a discussion about dopamine spikes or neural pathways. Our fears about smartphone and social media addiction reflect broad, longstanding fears about how technology impacts our norms and values, both as individuals and as a society. To understand change in this moment we must move beyond framing our problem with phones as a body problem, or a brain problem, or a self-control problem. This framing reinforces a dominant way of thinking in neoliberal capitalism: societal issues are collapsed into personal troubles that can be resolved through a series of microchoices.

Micro-solutions are certainly what tech companies are emphasizing today. Wary of getting the finger pointed at them, the tech titans are beginning to offer tools for self-monitoring. A recent Facebook ad implores users to "get outside," and both Facebook and Instagram have introduced time-tracking tools that offer "usage insights" to show users just how addicted they are. Google has embraced "digital well-being" with claims that it wants to help users get away from their screens; the newest Android OS, Android Pie, comes with an "app dashboard" that keeps track of exactly how much time you're wasting mouthing off on Twitter and how many times you pick up and unlock your phone. You can also download a "phone boss," an app that limits how much time you're allowed to spend on each app. YouTube will suggest "custom breathers," personalized reminders to take a break from viewing videos. Pinterest says it is working on encouraging users to "do," rather than view; it's building a "tried that" notification to encourage people to bake the cookies instead of just pinning a picture of them.

Sounds great. There's just one problem. These companies don't actually want you to put your phone down. Their business model depends on your spending ever more time posting, liking, searching, messaging, tweeting, self-monitoring.

More Than a Pretty Face

In the digital world, the old saying "She's more than a pretty face" takes on a new meaning. When you post that doe-faced close-up for you and your friends to see, tech companies see it too. They also see much, *much* more. They see that you snapped it in the Frog Pond rest room on the Boston Common. They see that you're hanging out with your cousin from Pennsylvania. That you two just had lunch at the Downtown Crossing Five Guys—the third time you've eaten at Five Guys this month. They know you're thinking about going on a diet, that you're starting to feel a little sad, that you're wondering if that's a lump in your armpit or your lymph nodes are just swollen—and that your cousin took a really long time in the bathroom! They know that later, after you got a cappuccino at Starbucks on the corner of Beacon and Charles, you hoofed it over to Uniqlo on Newbury Street, where you both looked at T-shirts. They see your message to your husband saying you're heading home in an hour and that there's leftover pasta in the fridge for the kids. They know you took a quick stroll through the Boston Public Library, then hopped on the Green Line. Settling in for a ride on the slowest train in the universe, they see you checked your email— that your best friend sent you a link to an apartment she's hoping to get and the couch she's thinking about buying if she does. They know that even though you told your boss you're psyched for the upcoming project, you're secretly scouring LinkedIn for a new gig because, as they see you told your mom the week before, you're worried the whole team will be let go after the current project finishes up. They see you bought your man headphones, that you two are planning to see a comedy show for his upcoming birthday, and that you spent the rest of the train ride on WhatsApp. And finally, that you got off near the Boston University Bridge and walked the rest of the way home.

These things that tech companies know about you are a mere taste, a tiny slice, of the information that is gleanable from one afternoon in the life of a moderate smartphone user. Jaron Lanier gives a sense of tech companies' omniscience:

> Algorithms gorge on data about you, every second. What kind of links do you click on? What videos do you watch all the way through? How quickly are you moving from one thing to the next? Where are you when you do these things? Who are you connecting with in person and online? What facial expressions do you make? How does your skin tone change in different situations? What were you doing just before you decided to buy something or not? Whether to vote or not?[37]

It's hard to overstate how much tech companies know about us. They are tracking us twenty-four hours a day, every day—even people who aren't connected to the internet, or are barely connected, are tracked. Anna Bernasek and D. T. Morgan, authors of *All You Can Pay: How Companies Use Our Data to Empty Our Wallets*, write, "When the database of what's known becomes sufficiently complete, what's left out can be deduced. Like a puzzle missing a piece, the contours of the 'hole' in the data show what's missing."[38]

Not all tech companies know an equal amount about us. The trouble is that we don't actually know how much we're being tracked, or how much which companies know about us, or what they are doing with the data they collect. Google probably knows the most, especially if you have an Android phone or a Gmail account. Tech sleuth Yasha Levine spells out what Google does with just some of the data it collects:

> All email communication was subject to analysis and parsed for meaning; names were matched to real identities and addresses using third-party data-bases as well as contact information stored in a user's Gmail address book; demographic and psychographic data, including social class, personality type, age, sex, personal income, and marital status, were extracted; email attachments were scraped for information; even a person's US residency status was established. All of this was then cross-referenced and combined with data collected through Google's search and browsing logs, as well as third-party data providers, and added to a user profile.[39]

Google recently stopped scanning people's emails for the purpose of selling ads, but it still scans and saves them, forever. Every search you've ever made, every bookmark, every download, every click, every Google Drive file (even those you deleted), every event (with time and location) on your Google Calendar, including whether you actually attended—is saved. If you have an Android phone, Google can access every single thing on it, including all the apps you use, your webcam, and your microphone. Even if you get a new phone the company saves all the information it collected from your old phone. Google also stores your location data. If you tap your Maps app, it will take a snapshot of where you were when you opened it. Any time you do a search it not only stores the contents of the search query but also where you were when you made a query: you typed "How do I know if I have insomnia?" while lying in bed. A recent Associated Press investigation whose results were confirmed by computer science researchers at Princeton found that even if you used Google's "privacy tools" on both Android and iPhones to ask Google not to track your location, it did so anyway.[40]

Your internet service providers—AT&T, Verizon, Comcast—also know a ton about you, which is not surprising, since they are the ones that provide the infrastructure to make your phone "smart." Salome Viljoen, a privacy expert at Harvard's Berkman Klein Center for Internet and Society, has warned:

> Your internet provider doesn't just know what you do on Facebook—it sees all the sites you visit and how much time you spend there. Your provider can see where you shop, what you watch on TV, where you choose to eat dinner,

what medical symptoms you search, where you apply for work, school, a mortgage. Everything that is unencrypted is fair game.[41]

Then of course, there's Facebook. Facebook has a detailed profile of every user, which includes but is not limited to basic identifying information (your name, gender, date of birth, ethnic affiliation), information about your networks (your employer, the clubs and associations you're affiliated with), details about how you access Facebook (your phone information, every IP address you've ever used to log in to Facebook, your operating system). The company also has a complete "activity log" that includes information on every action you've ever taken on any Facebook platform—including its WhatsApp, Instagram, and Messenger platforms—every post and photo, every "like" and click, even videos you never saved, messages you never sent, or friends you've deleted. If you're logged in to Facebook it tracks every other site you visit, and if you're logged out, it still tracks you through cookies and software innovations like Facebook Pixel. Facebook also buys data from data brokers such as Acxiom and Epsilon, which it combines with its own data to create an even fuller picture of you.[42]

Anytime you use your Facebook login to log in to another site, that company also gains access to your personal data: you've given access to your microphone, camera, photos, and contact list to Airbnb, Spotify, Amazon, Uber, Twitter, Pinterest, and all the other random apps smartphone users access. Researchers at Eurecom in France tested thousands of free Android apps in the Google Play store and found that many applications connected with hundreds of distinct URLs—including ad-related and suspicious sites—within just a few minutes of opening them.[43] All this data, drawn from what virtually everyone does while connected to the internet—increasingly collected through a machine that we carry everywhere with us—accumulates to an unimaginable degree. The photos, messages, comments, searches, and app downloads, and the metadata ("data about data"; what privacy expert Bruce Schneier calls "data smog"), we generate just by living are important.[43] The swipes and taps and easy slides into pockets and purses as we move along through our days and years adds up to something huge—a new frontier for global capitalism.

The Frontiers of Profit

Capitalism and frontiers are like peanut butter and jelly; they're linked together so often that we usually picture frontiers in terms of their subsumption into the ever-spreading for-profit system. When we think about how capitalism has spread slowly over the globe over the past five centuries, we imagine the frontiers that have brought new growth and change—the "New World," the "American West"—and we often associate these frontiers with new resources or new machines: silver mines, forests, railroads, steamships.

But the process by which frontiers are opened up is a bit hazy for most people. This haziness is partly a result of how the history of capitalism is taught—as an inevitable, inexorable process of transforming the world, a "natural" evolution. The haziness is also a result of the language we use. We "open" new frontiers the way we open a door. We imagine rubber trees and silver mines and rich soil just kind of sitting there, waiting to be transformed into a profitable venture. These dominant frames are not only wrong—they strangle our ability to comprehend the emergence of new frontiers and reinforce a determinism that locates change in technological advances, such as railroads or smartphones.[45]

To comprehend where we are headed in this technological moment we need to stand on firmer ground in our understanding of economic change. Frontiers are made, not opened. People make them through creativity, culture, knowledge, technology, blood, guts, power struggles, and lots of hard work. The creation of frontiers in capitalism has usually been a brutal, nasty process in which whole societies have been wrenched from their previous way of life and forced into new configurations. Grasping how capitalism evolves and expands—how new frontiers are made—requires putting people, rather than machines or geographical features, in the driver's seat.

So where are the people in capitalism? Most of the time we imagine them buying things, or running businesses and inventing things, or working on an assembly line or toiling in a mine. This isn't wrong. These people and the work they do are central to capitalist development. Equally important, though not usually theorized as such, is all the unpaid work that has gone into creating new frontiers in the history of our for-profit system—the appropriated work of slaves, of colonial subjects, of women.

We can't make sense of how capitalism has evolved without taking into account this appropriated unpaid labor. Only a small fraction of the work that goes into creating new frontiers is paid work. This is not a bug, it's a feature, as the saying goes. Sociologist Jason Moore, in his groundbreaking book *Capitalism in the Web of Life*, argues that capitalism depends on "cheap nature"—labor, resources, food, energy—and that the appropriation of unpaid work is (and always has been) as essential to the development of capitalism as paid work. "Absent massive streams of unpaid work/energy from the rest of nature—including that delivered by women—the costs of production would rise, and accumulation would slow."[46] In short, profit-making relies on massive amounts of appropriation.

Having lived with capitalism for some time now we know that it is neither judicious nor wise—it is voracious. It takes and takes until each new frontier is no longer a space that provides cheap or free resources; capitalist expansion diminishes the value of the forest or the mine to the point that it's no longer profitable to extract from it. Restless businesspeople and investors move on, using their capital to develop new frontiers, new places to make windfall profits—"new frontier[s] of appropriation."[47]

Today however, in the face of global warming, rising food prices, and habitat destruction, observers say we've reached a breaking point: like the old tree in Shel Silverstein's classic story *The Giving Tree*, the earth has given all that it can.[48] Increasingly, it appears that there are no new frontiers, or at least none with the potential for profit making and economic growth that frontiers have historically provided.

This may be true, but capital isn't listening. Silicon Valley has found a new frontier of appropriation and it's using your smartphone to "open" it. Instagram sold for $1 billion in 2012 despite only employing thirteen people at the time. WhatsApp had fifty employees when Facebook bought it for $19 billion in 2014. This is astonishing. Why are these companies worth so much? Business insiders say they are valuable because of their network potential. This is true, but also obfuscatory. Instagram's or WhatsApp's value, just like the value of so many other tech companies, is in the unpaid work they command, their ability to appropriate life—your life. On Facebook alone users collectively spend about two hundred million hours every day creating content. All of the tweets, selfies, reviews, homemade pet video compilations, stories, all of this life—because it *is* part of real life—is being appropriated by tech companies.

The appropriation of unpaid work isn't new. It is how frontiers are made in capitalism. In her pathbreaking book *Caliban and the Witch*, Silvia Federici talks about the long transition from feudalism to capitalism. Usually when we think about this transition we think about the historical process in which European peasants moved off (or were pushed off) of commonly held land and became wage laborers. But Federici talks about a parallel process in which not only the broader division of labor changed, but also the sexual division of labor. Men's work increasingly became waged work, and thus was officially valued in the emerging capitalist economy. Women's work, however, became devalued. It "began to appear as a natural resource available to all." This transformation encompassed life itself—life activities were redefined. New power dynamics that didn't exist before were created.[49]

This isn't a crude story about how women became oppressed and men became kings of their castles. Federici illuminates how women's unpaid labor became concealed, transforming the process of accumulation, and thus power relations for both men and women. The concealment of how important women's unpaid labor was to capitalism allowed *both* men and women to be oppressed: women were oppressed through the abuses and oppression of patriarchy and their manufactured dependence on male relatives as they were forced to toil unpaid in the home; men were oppressed through low wages because it was assumed that every man had a woman at home performing all the reproductive labor (cooking, cleaning, childrearing, household production, and so forth) necessary to maintain both him and society as a whole. Selma James, a feminist activist and scholar, neatly captures the appropriation of women's work: "The fact that it brought no wage had hidden that it was work. Serving men and children in wageless isolation had hidden that [women] were serving capital."[50]

James also writes, "The form of the relation between people through which the ruling class robs the exploited of their labor is unique in each historical epoch."[51] In the development of the digital frontier we are once again seeing a redefinition of life activities and the emergence of new power dynamics. Put differently, in the making of the digital frontier, we see how a new combination of appropriation and exploitation has been formulated, a model that has generated unimaginable wealth for the tech titans. Small numbers of highly paid software and hardware engineers alongside slightly larger numbers of support staff, blue-collar service workers, and partially hidden content managers are surrounded by a vast sea of unpaid users without whose unpaid work the digital frontier as it is currently configured would not be possible.

Once again we are witnessing the concealment of unpaid, appropriated work. Except today it's not just women's work that is being appropriated, being made to appear as a natural resource, a "labor of love." It is all of our work—the hours we spend every day on our smartphones creating content and generating data through our constant connection to our hand machines. In these hours, our lives become ever more deeply enmeshed in the circuits of capital. Our appropriated work, and our digital selves more broadly, are the key to the digital frontier. We generate the data on which this frontier rests.

"Big data" is an inelegant yet apt moniker for the massive increase in the volume, variety, and velocity of data generated in the past decade. With the extension of the internet to more spheres of life, combined with the dramatically increased processing power and storage capacity of computers, and the democratization of machine learning through easy-to-use open-source software packages, we are generating, saving, and processing an unfathomably large amount of data.[52] The data is so vast that we resort to made-up-sounding words to describe and predict it—exabytes, zettabytes, yottabytes.

The quantity of the data, much more than its quality or the algorithms used to process it (though many are quite powerful), has many observers seeing an endless horizon of profitability and innovation. Data is now valuable. What was once viewed as frivolous and useless has become valuable. Two big-data experts, Victor Mayer-Schönberger and Kenneth Cukier, argue: "Although data has long been valuable, it was either seen as ancillary to the core operations of running a business, or limited to relatively narrow categories such as intellectual property or personal information. In contrast, in the age of big data, all data will be regarded as valuable, in and of itself."[53] Levine likens it to "a new form of alchemy" in which Google and other big tech companies have transformed "useless scraps of data into mountains of gold."[54]

Big tech companies and third-party data vendors are amassing, hoarding, trading, selling, and extracting data like gold. As mentioned earlier, Google and Facebook already scoop up all the data we generate on their platforms and much of the data we generate off them, appropriating our rich, dynamic digital lives. They develop and use algorithms to collect and sort this data for targeted advertising. Facebook offers advertisers nearly a hundred known targeting options: the size of a user's home, their ethnic affinity, their educational attainment, whether they are expecting a child, whether they are conservative or liberal, what style and brand of car they drive, to name just a few. In 2018, Facebook made $55.8 billion in revenue, $55 billion of which came from advertising; 92 percent of its total ad revenue came from exploiting the data we generate 24/7 with our pocket computers.[55] Google made roughly $115 billion in ad revenue in 2018—about 70 percent of its parent company Alphabet's total revenue. Google is not quite as dependent on mobile advertising as Facebook, but in 2017 it earned 67 percent of its ad revenue from the devices we carry everywhere with us.[56]

The data we generate just by living and socializing is also collected and hawked by data brokers. Data brokers have thousands of pieces of information on virtually every person in the United States. Acxiom is a San Francisco company offering "identity resolution services." In 2018 it offered up to 10,000 attributes on each of the 2.5 billion "addressable consumers" (from sixty-two countries) in its database.[57] They slot each of us into categories on the basis of race and ethnicity, marital status and family size, health, gender, and economic status and then sell this information to companies. We don't hear much about these data brokers; they come into the spotlight only when someone bothers to go sniffing. In 2013 Senator Jay Rockefeller of West Virginia held hearings that revealed why data brokers might prefer to stay under the radar: one broker, Medbase200, a data broker based in Lake Forest, Illinois, peddled lists of "rape sufferers" ($79 for 1,000 names), domestic violence victims, HIV/AIDS patients, and "peer pressure sufferers" to pharmaceutical companies.[58]

Computer algorithms have been used to lump people into categories for decades. In the late seventies, Jonathan Robbin, a social scientist and computer expert, developed a software package called PRIZM that used zip codes and census data to sort Americans into forty marketing "clusters," such as "Shotguns and Pickups" and "Young Suburbia."[59] Today, as a result of our umbilical connection to our smartphones, there is infinitely more data, and opportunities to trade and sell and steal this data. But marketers still slot us into colorfully named consumer categories: "Rural and Barely Making It," "Tough Start: Young Single Parents," "Rough Retirement: Small Town and Rural Seniors," and "Zero Mobility" sound a lot like Robbin's originals.[60]

Once slotted, we're also scored. Companies want to know whether we can be trusted to take our meds, pay our bills, keep a job, stay healthy, buy stuff, and so on, so they use predictive modeling—utilizing thousands of individual factors or data streams—to assign us consumer scores. According to Pam Dixon, executive director of the World Privacy Forum, and the privacy expert Robert Gellman, hundreds, maybe thousands of consumer scores exist: credit scores, health risk scores, energy consumption scores, propensity for pregnancy scores, job security scores, medication adherence scores, churn scores (the likelihood that you'll move your bank, phone, cable, etc., business to another merchant).[61]

Virtually every adult in America has at least one consumer score (and in all likelihood many, many more) attached to them. Yet, the details of what scores are attached to whom, what institutions use these scores, and how they use them are fuzzy, to say the least, because most of these scores are secret and unregulated. A *Wall Street Journal* story suggested that how long you are put on hold when you call customer service might actually be determined by your "customer lifetime value score"—essentially how much money you spend. A low customer value score will have you on hold for a long time.[62]

Financial institutions, law enforcement, businesses, and healthcare providers use a variety of scores to predict behavior and make eligibility decisions, but ordinary people aren't able to see, correct, or control these scores. Moreover, according to Dixon and Gellman's research, the companies that construct consumer scores to sell to health insurance companies, retailers, and other businesses don't follow federal guidelines related to credit scores. Specifically, score generators include factors such as gender, race, marital status, religion, and national origin in the construction of scores, which is illegal according to the Equal Credit Opportunity Act when constructing credit scores.[63]

Categorizing and scoring us is just the beginning. Retailers dream of a future where they can follow us around and show us what we want to buy, or coerce us into buying something at any given moment. Anindya Ghose, a professor of business at New York University, breathlessly describes turning one's smartphone into a "pocket butler": "If you, the consumer, let marketers learn your habits and read your mind, you will let that smartphone become your concierge and save you time."[64] In this shopper's paradise, a combination of beacons, geocoding, Wi-Fi, RFID, near-field communication, and other digital identification and communication technologies will tell companies where we are at all times, and leave us open for suggestions and offers. iBeacon and Google's Eddystone for Android phones show that Ghose's fantasy is fast becoming a reality.

Fintech companies—businesses that use modern technology and algorithms to compete with traditional financial service providers—are equally enthusiastic about capitalizing on the data we continuously generate with our smartphones. Admiral, a small UK start-up, hoped to mine Facebook data to create a new formula for generating credit scores. Specifically it wanted to read people's Facebook posts and on the basis of the wording determine what people's credit scores should be. "The use of exclamation marks, for example, could suggest that the writer is overconfident. Lists, on the other hand, could signify a more organized, cautious person." Sounds kooky, but Admiral isn't alone. Kreditech, a fintech start-up in Hamburg,

Germany, caters to people with little or no credit history, offering them loans, checking accounts, and so forth based on applicants' voluntarily submitted tweets and Facebook posts.[65]

Facebook shut down Admiral's party before it could get started, but not because it thought it was a bad idea. Facebook filed its own, slightly different, patent so it could sell data to lenders rather than have third parties such as Admiral scrape it. Facebook's patent application read: "When an individual applies for a loan, the lender examines the credit ratings of members of the individual's social network who are connected to the individual through authorized nodes. . . . If the average credit rating of these members is at least a minimum credit score, the lender continues to process the loan application. Otherwise, the loan application is rejected."[66] The implications of Facebook's patent are disturbing, foreshadowing a future in which our ability to get a mortgage may be determined by whether our friends pay their bills on time—a future in which people are deemed uncreditworthy by association.

Next in line to mine our personal data are the insurance companies. They imagine a world in which risk can be "measured in feet and inches," where people's every move can be tracked and, if possible, insured. Instead of paying a yearly fee with premiums only increasing after an accident or decreasing as a result of repeated safe behavior, insurance companies envision real-time flexibility:

> Imagine that mobile phone signals . . . detect that a person is about to walk down a road where several people have recently fallen on ice. The insurer will react by either sending a message warning the person to walk more carefully or else automatically increase the premium and coverage while the policyholder is walking down that road.[67]

Granted, not all uses of big data are designed to sell things. There is much new knowledge that can be gleaned. Efficiencies can be gained. Google sister-company DeepMind, for example, uses electricity usage data to reduce waste. Big data can also be used to predict weather patterns, improve crop yields, and develop new drugs. But the vast majority of current big data use revolves around selling digital selves for profit, and the well of data seems infinite. Leading up to Facebook's IPO market researchers estimated that between 2009 and 2011 alone the company had collected more than two trillion pieces of "monetizable content."

Silent Partner

When we think about the creation of frontiers in the history of capitalism we imagine the companies and the men who led and achieved vast wealth from these endeavors. Less emphasized is the central role of governments in making capitalist frontiers.

The Union Pacific Railroad, America's first transcontinental railroad, which connected the East and West Coasts, was funded by US government bonds approved in a series of congressional acts beginning with the 1862 Pacific Railroad Act. The railroad companies involved in the project received 175 million acres of public land between 1850 and 1871. Also to the benefit of companies and settlers, the US military cleared the land of Native Americans—a population that the historian Alan Trachtenberg argues was seen as "the utmost antithesis to an America dedicated to productivity, profit, and private property."[68] In short, private property and corporate rights in the new western frontier were enforced by a deep partnership between US companies, the federal government, and the military.

The intertwining of martial virtues and entrepreneurial adventurism has never wavered in the development of the United States, but it slipped from public attention in recent years. Until, that is, Edward Snowden, a system administrator for government consulting firm Booz Allen Hamilton, fled to Hong Kong in 2013, taking a thumb drive containing thousands of classified NSA (National Security Agency) documents with him. The US government is usually seen as a bit player in the modern digital frontier. Although many are aware that the first version of the internet, ARPANET (Advanced Research Projects Agency Network), was created by the Department of Defense in the late 1960s, most people familiar with the internet's creation assume that when the government gave the internet to private telecoms in the early nineties it reduced its role in the digital world. Snowden revealed that this perception couldn't be further from the truth. Like the transcontinental railroad and the conquering of the West, the digital frontier is being made through a deep partnership between Silicon Valley and the US government.

This partnership takes myriad forms. For example, tech titans marketize technologies developed by the military; according to the technology writer Peter Nowak it is nearly "impossible to separate any American-made technology from the American military."[69] The US government uses its influence in international institutions and its nearly eight hundred military bases abroad to open doors for US tech companies globally. As Robert McChesney notes, "The government is like a private police force for the

Internet giants."[70] Tech titans also have ongoing and extremely lucrative tie-ups with US intelligence agencies: Google and Palantir (founded by the PayPal billionaire Peter Thiel) have deals with the CIA and the NSA, and the CIA is one of Amazon's best customers; in 2013 it signed a ten-year, $600 million dollar cloud computing contract with the e-commerce giant.

More important, just like the tech titans, the US government tracks, records, and stores every moment of our digital existence. The impulse to spy on Americans is not new. "Military and intelligence agencies used the network technology to spy on Americans in the first version of the Internet," Levine contends. In *Surveillance Valley* Levine details how endeavors such as ARPA's Project Agile in Vietnam demonstrate how one of the foundational drives in creating the internet was the development of technology to spy on civilians: "Surveillance was baked in from the very beginning."[71] As storage has gotten cheaper and technology more advanced, the US government has, like private tech companies, massively increased the amount of data it collects and stores. In 2014, the US intelligence community, led by the NSA, finished construction on the Utah Data Center (a.k.a. Bumblehive), a massive cloud storage facility at Camp Williams near Salt Lake City.

The former chief technology officer for the CIA, Ira "Gus" Hunt, summed up the government's philosophy when he said, "We try to collect everything and hang on to it forever."[72] Through his disclosures about NSA spying, particularly the NSA's PRISM program (given the green light by the Patriot Act), Snowden showed Americans how they are subject to what Schneier calls "ubiquitous mass surveillance."[73] The NSA collects internet communications from Microsoft, Yahoo, Google, Facebook, YouTube, AOL, Skype, Apple, and many other tech companies whose services millions of Americans rely on. Meanwhile, the CIA's Mobile Devices Branch has perfected a wide range of hacking systems, malware, and viruses that it uses to track people through their smartphones. Documents from WikiLeaks show that both the NSA and the CIA can bypass encryption.[74]

"Corporate and government surveillance interests have converged," Schneier contends. "Both now want to know everything about everyone. The motivations are different but the methodologies are the same."[75] The digital frontier rests on this convergence. Both companies and the government are appropriating our unpaid work, our digital selves—the life we live through our phones, all the content we produce and data we generate—in order to create a new frontier of profit making and control that seems limitless and beyond our ability to control.

The Cloud's the Limit

The notion of the infinite is pervasive in the digital frontier and has echoes in past frontiers as well. When colonists landed on the shores of the American continent in the early 1600s, they saw its forests and imagined an infinite number of trees. Thomas Jefferson said, "It will take a thousand years for the frontier to reach the Pacific." They were wrong, of course—settlers had ravaged the country's forests by the middle of the nineteenth century. Frontiers always have limits, and the digital frontier is no exception.

The first limit of the digital frontier is an ecological one. "Cyberspace" conveys a sense of boundlessness; we imagine ourselves as nodes in a vast unknowable network sending packets of information through the cloud—a fluffy white puff of air. Tung-Hui Hu, a former network engineer who is now an English professor, argues masterfully, "The term cloud refers to the same cultural fantasy of its analog namesake," the idea of the air as "inexhaustible, limitless, invisible" and the cloud as "a reserve of seemingly unlimited computing power, or storage space." Hu says the "cloudlike nature of the network has much less to do with its structural or technological properties than [with] the way we perceive and understand it; seen properly the cloud resides within us."[76]

We are attached to a fantasy about the cloud that is radically divorced from reality. The hardware and cables and networks that make the internet go are physical, they exist in real spaces—computers, buildings, and data centers that are often hidden from the public eye. These physical elements of digital life require energy to produce and run, vast amounts of energy. In a recent report, Lotfi Belkhir and Ahmed Elmaligi, researchers at McMaster University in Toronto, predict that the rapid growth of the information and communication technologies (ICT) sector, which comprises data centers, desktops, laptops, displays, computer hardware, and smartphones, will increase this sector's greenhouse emissions to 14 percent of global emissions by 2020. Data centers, many of which are coal-powered, are by far the biggest source of energy consumption in the ICT sector, but smartphones themselves, the authors show, are also quite toxic. A colossal amount of effort and energy is required to extract the rare metals needed for each phone. Yet, as a result of built-in obsolescence, we only keep them for about two years before getting a new one; production accounts for 85–95 percent of the smartphone's total carbon footprint.[77]

Our fantasies about the digital frontier hide the hierarchical and ecologically destructive relationships of global capitalism. We see neither the people who toil in terrible conditions making our phones, nor the broader production chain that begins with exploitative and destructive extraction of rare metals and ends with toxic e-waste dumped in poor countries and communities. In the swirl of ideas surrounding the digital frontier, we're encouraged to keep our thoughts focused on how many social media followers we have or what we're going to buy and get delivered in record time.

Our smartphones have reinforced our role as consumers. As we tap our screens for an Uber, the driver arriving ever faster, we forget that there are other, more sustainable, types of transport. When we activate our Amazon app to order some small convenience, which arrives in layers of packaging like a matryoshka doll, we elide the amount of waste we produce through our unreflexive consumption. Moreover, the experience economy has a sense of weightlessness about it, as if we were only consuming fun or ideas. It hides how the world of social media has been designed to sell advertising to get us to buy real things. Overall, the digital frontier is characterized by exponentially growing energy and resource consumption.

There's a sense that data, when there is enough of it, takes on a magical quality. Mayer-Schönberger and Cukier marvel, "In the big-data age, data is like a magical diamond mine that keeps on giving long after its principal value has been tapped."[78] In this vision, data is infinitely useful and durable. But all data, even big data, has a shelf life—an economic limit to the digital frontier—after which its profitability is radically diminished. The data that is gathered about us, in particular about our consumption habits, is never enough. Companies need fresh data, more data.

The primary use for big data is advertising. But there's only so much advertising each consumer can absorb; after all, we don't have infinite dollars to spend or attention to spare. People get good at spotting and ignoring ads. Schneier contends that right now

the value of a single Internet advertisement is dropping rapidly, even as the cost of Internet advertising as a whole is rising. Accordingly, the value of our data to advertisers has been falling rapidly. A few years ago, a detailed consumer profile was valuable; now so many companies and data brokers have the data that it's a common commodity. . . . This is why companies like Google and Facebook keep raising the ante. They need more and more data about us to sell to advertisers and thereby differentiate themselves from the competition.[79]

These limits are why tech companies are constantly trying to collect fresh data. Some of this expanded data collection involves increasing the scope of surveillance. For example, Uber tried to get riders to note which friend they were going to see instead of the address to which they were headed, and increased the time it would track each passenger to five minutes after the ride ended so the company could see and note where the passenger walked, what store or business they entered, and so forth. (It stopped doing this after customers complained.) Companies also get new data by moving to new demographics. Facebook is growing both up and down: older populations have increased their use of the social media platform dramatically and recently the company developed Messenger Kids, appealing to parents to set up accounts for children as young as twelve. Google has hooked young users through its massive, years-long push into public schools, ensuring that students are enmeshed in the Google ecosystem from a young age.[80]

But the richest source of new data is getting people to "share" more. Photographer Martin Adolfsson and artist Daniel J. Wilson created an app called Minutiae; the app is styled as a game and sends users "push" notifications at random points every day asking them to take a picture of something around them. "Minutiae is your automated self-portrait. Real life. Uncurated. Unfiltered. Unfollowed," its creators say. It's also a novel way to generate fresh data on the interstices of life.[81] Twitter's increased character limit; Facebook Live, which kicked off with celebrities like Kevin Hart and Michael Phelps livestreaming for fans; Amazon's voice-activated Echo Look device, which takes short videos of users trying on outfits in their bedrooms—are examples of the drive for ever more sharing; all create vast new streams of data and information.

Viewing the digital frontier as a development in a historical system that has relied on appropriating cheap or free nature and unpaid and underpaid work shows us social limits to the digital frontier in addition to economic and ecological limits. In the past, things that were once considered free forever, always there for the taking, became dearer. The ease of appropriation ended, forcing capital to seek new directions for profit making.

Women's unpaid work is a prime example. By the nineteenth century the long struggle to keep women inside the home seemed successful, and the image of the "natural" role for women, the content, happy mother ruling over her domestic environs, was complete.[82] (It was never entirely successful because poor women always found a way to engage in paid work outside the home in order to feed their families.) This image, however, papered over a simmering discontent that erupted in waves of feminist uprising. In the 1970s women took to the streets demanding an end to patriarchy. They formed a campaign provocatively called Wages for Housework that demanded recognition for all the essential labor they performed for no pay. The struggle is ongoing, but as women have entered the waged workforce en masse over the past four decades, norms have slowly shifted. No longer do women expect to get married and remain at home; more important, no longer do men and society at large expect women to be saddled with providing all of the labor necessary for the reproduction of the work force. These shifting norms demonstrate the contingency of the cheap nature we take for granted. Things that were once free for the taking can become unavailable, or no longer free.

The exposure of women's concealed labor helps us imagine the social limits of the digital frontier. Right now we accept big tech's bargain. We get cool apps and tools to communicate with others and to entertain and educate ourselves. Companies get unlimited access to and control over all the data we generate with our perpetually connected hand machines. This bargain is tenuous, however. We're increasingly uncomfortable with the relationship we've developed with our smartphones, uneasy with the ways we interact and express ourselves in our phone worlds, and fearful that our increasing dependence on our smartphones will overpower our fragile sense of authenticity and self. Discussions about how we use our phones are laced with loathing and judgment, aimed at ourselves and others. We blame ourselves for being weak and narcissistic.

To a degree we are weak and narcissistic. But we should be wary of explanations that blame individuals for an issue that an entire society struggles with. As more and more people become suspicious of the technology, institutions, and relationships embodied in their phones they are taking a closer look at their hand machines and the companies who control them. Our fears express a growing awareness of our vulnerability vis-à-vis the tech giants—a growing sense that life itself is somehow being shaped around the needs of profit making.

So while we should absolutely be much more thoughtful about what we do on our phones and why, we should also take a harder look at the new digital frontier of profit making. Few of us have profit making in mind when we send a message to family and friends on Facebook or post a picture of the amazing tiramisu we're about to scarf down. Yet the digital frontier is rooted in appropriating these very activities—turning life itself into a moneymaking venture. This isn't a notion most of us are in favor of. We relish free spaces outside the sphere of the market. Our family, friends, communities, and leisure time have long been places of refuge away from the market. Today these free spaces are threatened. Our phones are Trojan horses used by tech companies to capture these spaces—to turn them into spaces of profit making. Recognizing this allows us to reimagine our phones as an important site of struggle in the forging of the new digital frontier. However, our hand machines are also a *tool* of struggle.

Chapter 4

New Politics

President Trump loves to tweet.

He tweets when he's mad:

> Sorry losers and haters, but my IQ is one of the highest—and you all know it! Please don't feel so stupid or insecure, it's not your fault.[1]

He tweets when he's sad:

> Why would Kim Jong-un insult me by calling me 'old,' when I would NEVER call him 'short and fat?' Oh well, I try so hard to be his friend— and maybe someday that will happen![2]

He even tweets when he's glad:

> Happy #CincoDeMayo! The best taco bowls are made in Trump Tower Grill. I love Hispanics![3]

Trump's tweets are a hallmark of his presidency and may even hold the key to how he managed to pull down the win when even his own polls showed him losing on Election Day. Trump's 2016 campaign behavior seemed like political suicide. His attacks on immigrants, women, fellow candidates, and media personalities revealed a vindictiveness and vanity that public figures are usually eager to conceal, a careless ignorance and mendacity that would derail most mortals. Yet Trump's tweets didn't cause campaign combustion. Instead, they revealed a stark truth. The reality TV star had a bead on what much of the country wanted to hear: "Build a wall." To a large portion of the electorate Trump's micromissives, read on smartphones all over the country, felt authentic and purposeful.

This isn't the first time an American president has connected with the masses through a new technology. The stories we tell schoolchildren about Franklin Delano Roosevelt often include a reference to his "fireside chats," which made a country in the throes of economic depression and political uncertainty feel cared for and secure. John F. Kennedy changed political campaigning forever when he appeared in a televised debate with Vice President Richard Nixon in September 1960. Americans were impressed with how handsome and presidential Kennedy looked compared with Nixon, who had just come down with a bug and appeared sweaty and pale in the telecast. Despite presenting a broadly similar quality of responses, Kennedy's favorable screen impression put him on top, helping pave the way to his victory—and to a new kind of politics rooted more deeply in style rather than substance.

Technology is once again at the center of politics. In the most obvious sense, our attachment to our phones and social media has made it normal now for the president to address the nation by dashing off a tweet (or likely having Dan Scavino, the White House director of social media, do it), and for people to read this message while sitting on the can or in the car, and to post a reply if they wish. But more than just catalyzing a shift in how people absorb or express political sentiments, our smartphones are at the heart of a new political moment.

The Political Is Personal

One of the most famous rallying slogans to come out of the late 1960s feminist movement was "The personal is political." The phrase had divergent interpretations, but it broadly emphasized the connection between women's personal trials and tribulations (controlling partners, domestic drudgery) and the broader political structures and processes of a for-profit system rooted in sexism, racism, and the oppression of working people. Today, the notion that the personal is political continues to resonate. It captures the struggles of ordinary people, especially women, people of color, and millennials, to find work-life balance, to manage debt, and stave off alienation in neoliberal capitalism, an economic system that champions competition, shareholder value, and the slow creep of market relations into every sphere of life.

But we can also think about the reverse today—the political has become personal. With our smartphones always at the ready, politics has become seamlessly interwoven into our daily lives: on the way home from work we'll comment on a tweet thread about a teacher strike happening hundreds of miles away; drinking our morning coffee, we'll share a video of students sitting in at their congressperson's office; waiting for the bus we'll mark "attending" for a Facebook event featuring a local woman running for city council.

Smitha Chadaga, an Oregon doctor, is someone for whom politics has become personal. When her six-year-old son expressed worries that he'd have to leave the country if Trump got elected, Chadaga decided she needed to become more politically active. She joined Physician Women for Democratic Principles, and in the run-up to the 2018 midterm elections she began reaching out to fellow voters through mass texting apps on her phone. In free moments throughout her day, at lunchtime, working out at the gym, whenever, she used the apps to reach out to voters: "You send 50 texts in two minutes. So if I get to the school to pick my kids up early, I can send 50 texts."[4]

Politics has always found its way into our personal lives, but smartphones combined with social media are at the center of a shifting political landscape. We see marked changes in how people participate in politics. Social media and online news aggregators have become a major, and in some cases primary, source of news. This switch is coupled with the pre-smartphone emergence of the twenty-four-hour news cycle; together these shifts have upended our expectations about getting "the news." Instead of relying on nightly news broadcasts, or a daily newspaper delivered to our door or picked up with our morning coffee, we get instant, regularly updated news about nearly any topic with the tap of an app. If we hear that a fire has broken out nearby, we'll scour Twitter, confident that someone witnessing the event is playing journalist. If a tragedy happens we expect updates on the hour from multiple news sources. We demand information.

But politics is much more than reading or listening to the news. Politics is a dynamic relationship in which people and institutions from above and below create, reproduce, enforce, and change the rules, structures, and norms of society. With our phones in hand, we're doing much more than obsessively scrolling through our news feeds. We're transforming the way politics is done.

In November 2014, Synead Nichols was furious. Darren Wilson, the cop who had gunned down Michael Brown Jr. in Ferguson, Missouri, in August of that year was found innocent by a grand jury. Nichols later told *Essence* magazine, "I kept thinking, what do I do? I said, You know what?

Facebook. Arab Spring did it, Brazil did it, Mexico did it, so why can't New York do it?" Nichols made a Facebook group and added her friend Umaara Elliot as a host. A few weeks later the two young women marched at the head of a fifty-thousand-strong rally in New York City that they had organized to demand justice for those wrongfully killed by police.[5]

Two years later, on the other side of the world in Aleppo, Syria, six-year-old Bana al-Abed and her mother, Fatemah, decided the world needed to know about the bombs raining down destruction around them. Little Bana sent out a tweet on September 24, 2016: "I need peace." Those three words started a global conversation. Charging their phone with a solar panel and picking up whatever mobile phone signal was available, Bana would tell her mother what she was thinking or feeling and her mom would tell the world in a tweet. More than all the news coverage, Bana's voice brought home the horror of the Syrian war. Before finally escaping to Turkey, Bana had gained millions of followers from around the world.[6]

In February 2018, a former student who had bragged on social media about wanting to shoot up a school walked into Marjory Stoneman Douglas High School in Parkland, Florida, armed with an AR15 rifle, pulled a fire alarm, and gunned down seventeen students and staff members. Safe in the car with his dad and brother later that day, MSD student Cameron Kasky pulled out his phone to talk about what happened . . . and didn't stop. The next day Kasky and his friends stayed up all night creating new social media accounts for a new social movement—#NeverAgain. The students used their own networks and created new networks to organize a march on Washington and push for legislation in the Florida statehouse to strengthen background checks and increase the minimum age to purchase a gun.[7]

Critics say digital technology is atomizing and individualizing, yet the ways that phones are making politics personal challenge this interpretation. Our mobile connection has not only fostered the sense that people can raise their voices and be heard but that they should. Instead of waiting for someone to tell us what to do—to get out into the streets, or organize an event, or speak up about something—our hand machines have encouraged us to take matters into our own hands.

People figured out that mobile phones could be repurposed for other-than-intended uses pretty much instantly, and that includes politics. Joseph Estrada, the Philippine president, was ousted in 2001 when thousands of protesters gathered in central Manila in response to a simple text they'd received on their phones, "EDSA," the initials of Epifanio de los Santos Avenue, the location of the planned protest. Pre-smartphone-era protests against the World Trade Organization, actions outside the 2004 Republican Convention, and many others worldwide have been organized using mobile phones.[8]

The internet, too, has long been a space for sharing political ideas, and for organizing and networking. Alter-globalization activists who participated in the anti–Free Trade Agreement of the Americas actions held in Quebec City in 2001 used the web to link up with other activists from across North America to find food, lodging, and legal aid. MoveOn.org, an organization started by two tech entrepreneurs, Joan Blades and Wes Boyd, during the 1998 impeachment trials of President Bill Clinton, laid the groundwork for the features of many modern political campaigns, including online petitions, crowdsourcing donations, virtual phone banks, and links between digital and analog organizing.

The smartphone–social media connection has built on this history. But smartphone politics has also catalyzed something new. Our constant digital connection and access to vast networks facilitate new ways of doing politics. For one thing, our phones allow us to bypass mainstream media, which is often selective in its coverage. A conversation between social movement journalist Sarah Jaffe and Rasheen Aldridge, a participant in the Ferguson protests following the killing of Michael Brown Jr., captures this dynamic: Initially Aldridge said he didn't pay much attention to accounts that a young man had been shot in St. Louis because it was a depressingly common occurrence. "But once I went on Twitter, I saw the details and the response of the community people reporting, instead of the news reporting on it; [I] got a different idea of what was actually going on."[9]

Digital movements also have the potential to "go viral," particularly when celebrities or corporations get involved. When George Clooney and Oprah pledged allegiance to the #NeverAgain cause, the Parkland students received an outpouring of support from other celebrities and corporations who promised to make changes. Bumble said it would ban all photos with guns from its dating profiles, while Dick's Sporting Goods pulled all modern sporting rifles from its shelves and destroyed them. Virality, or unpredictability more generally, is a persistent feature of political movements. As world-systems sociologist Janet Abu Lughod noted, "Large disturbances sometimes flutter to an end while minor ones may occasionally amplify wildly, depending upon what is happening in the rest of the system."[10] Our hand machines with their always-on digital networks heighten the potential for amplification.

Modern-day social movements, with one foot in the cloud and one foot on the ground, foster a deep sense of solidarity and participation for people who are often far from the political action. Tech and media expert Zeynep Tufekci says, "Especially in real-time situations, it is as if social media create an umbrella that envelops the protest and at the same time reaches out to people, potentially millions, who feel that they are part of the movement."[11] This was certainly the case as the world watched the 2011 protest in Tahrir Square unfold. Thousands of protesters gathered in Cairo demanding the resignation of Egypt's repressive leader, Hosni Mubarak. Standing firm in the face of a violent state response, the protesters publicized their campaign to the world in real time. They gave live interviews to the BBC on their phones, posted YouTube videos of government violence, and communicated with supporters around the world. This communication conveyed the drama unfolding and fostered such a strong sense of solidarity that one supporter, a thousand miles away in a Gulf state, started @TahrirSupplies to coordinate global donations of medical supplies.[12]

Some of the savviest people doing smartphone politics are not activists like those in Tahrir Square or Parkland High but are the people in power themselves. Smartphones have woven politics more seamlessly into our lives, and as a result we see and hear the voice of political figures more often as they use social media to connect with and influence constituencies. And politicians are expected to foster this digital relationship. Rather than appearing periodically on the campaign trail or at planned speaking events, politicians pepper us with regular updates about both their political and personal lives. We can mark a politician's reach and influence by their follower count on Twitter. Former president Barack Obama is the clear winner with 102 million Twitter followers. The scramble below can get weird. Former congressman and 2020 Democratic presidential candidate Beto O'Rourke, who tried and failed to unseat Republican senator Ted Cruz in 2018, recently broadcast himself on Instagram Stories getting his teeth cleaned—holding the phone up to give Americans a view of his gaping mouth—to publicize the situation at America's southern border.

President Trump has fifty-five million Twitter followers and calls his use of social media "modern day presidential." News of national import, such as Trump's decision to ban transgender soldiers from the US military, and personal gripes, such as his opinion that NBA star LeBron James is dumb, both find their way to the American public through Twitter. The nature of Twitter means that ordinary Americans can respond to or share these announcements with any amount of candor they wish. In May 2017 Holly Figueroa O'Reilly, a Seattle songwriter, replied to a Trump tweet with a popular gif of Pope Francis rolling his eyes at the president and the comment "This is pretty much how the whole world sees you."[13]

Trump responded to Figueroa O'Reilly the way many Twitter users do to someone who disrespects them—he blocked her. Trump has blocked hundreds of Americans from viewing or commenting on his page. Trump's Twitter blocks raise an interesting issue for politics in our smartphone society. Trump's page is not his personal page. It is, former White House press secretary Sean Spicer pointed out, "the official mouthpiece of the US president." If Trump blocks Americans from responding to or viewing his tweets, he is effectively blocking their First Amendment rights. At least this was the argument that the Knight First Amendment Institute at Columbia University made when it sued Trump in a class-action lawsuit. The judge agreed with the plaintiffs and ordered the president to unblock them. According to the ruling, if Trump wants to use Twitter in his official capacity as president he'll have to put up with snark.[14]

Outside the United States, social media platforms play a similarly important role. In India, where people primarily access the internet through their phones, Prime Minister Narendra Modi relies heavily on social media to connect with voters. For Modi and many other politicians, Facebook offers something that the old MoveOn-style campaign tactics and political blogs could never dream of—the ability to target and tailor messages for small groups of voters clustered according to characteristics such as income, educational attainment, gender, religion, ethnicity, or animosity toward minorities. Modi, who has more than forty-seven million Twitter followers and his own app, has created a "digital army" to spread the word of the Bharatiya Janata Party, India's ruling party. Sadhavi Khosla, a BJP cyber volunteer for two years before she quit, supported the candidate by trolling dissenting voices on social media. Khosla, who eventually became disturbed by the misogynistic, Islamophobic messages she was asked to disseminate, was part of an organized network of BJP trolls who used social media platforms to attack and delegitimize critics of Modi online.[15]

Political parties are also hiring politically neutral tech companies such as NationBuilder to build social media campaigns for them. NationBuilder matches voter data with social media profiles to raise money and fire up volunteers. Jacinda Ardern, New Zealand's youngest female prime minister; La République En Marche, led by Emmanuel Macron; Jagmeet Singh, the first person of color to lead a national party in Canada; and the Maryland GOP—all used NationBuilder to channel their social media presence into political victory. The company's late cofounder, Jim Gilliam, saw the company's role as "democratizing democracy": its sliding-scale pricing model makes it a popular choice for smaller campaigns and local candidates, such as "No Money Mike" Connolly, a progressive candidate from Cambridge, Massachusetts, known for eschewing corporate donations, who now represents the Twenty-Sixth Middlesex District in the Massachusetts House of Representatives.[16]

Our connectedness and willingness to make politics personal has also made it easier for governments to use our phones to crack down on us. Governments not only track people using their phones, but also attempt to curtail their ability to exercise their political voice and build political movements through their phones. Often governments do this by simply turning off mobile networks or the internet itself. This is how protests in Nepal in 2006 and Myanmar in 2007 were dealt with.[17] In 2011 President Hosni Mubarak shut down the internet for six days, leaving only the NOOR Data Network (used by the Egyptian stock exchange) functional. In late 2018 ruling party officials in the Democratic Republic of Congo shut off the entire internet for twenty days following a contested presidential election. But as economies become more reliant on digital connection, and activists become more savvy, governments are opting for more sophisticated strategies.

In the People's Republic of China, state officials use a combination of algorithms, government directives, and people power to control the digital political scene. With a population of nearly seven hundred million smartphone users this is a daunting task, but the party leadership has made it a priority. Recently, for example, Sina Weibo, China's largest microblogging platform, announced that it had given government officials access to its posts, enabling officials to tag certain posts as rumors. The Weibo incursion is one example of a broader push in China, led by the country's internet tsar, Zhuang Rongwen, to combat "wrong ideological trends."[18]

Another example of this effort is China's "social credit system," in which people are given scores based on their transaction history, habits and lifestyle choices, social networks, and their record of fulfilling contractual obligations. Being politically active, for example, will lower one's social credit score. The dissident journalist Liu Hu has been deemed "untrustworthy" and tarred with a social credit so low that he is banned from purchasing plane or train tickets, taking out a loan, or using social media.[19]

China also has one of the strictest internet censorship regimes in the world. In 2017, images of Winnie the Pooh were banned from social media after users compared Xi Jinping, the president of China, to the honey-loving bear. Apps like TikTok, a short-video app, will now be held responsible for content that violates any of the country's one hundred types of "inappropriate" content.[20] The People's Republic doesn't just administer the "great firewall of China"; recently a man who compared Xi to a "steamed bun" on a private chat app was sentenced to two years in prison. Using apps such as Lantern to bypass China's firewall and gain access to banned websites and apps now brings fines and jail time.

Other countries have followed suit. In the fall of 2018 Saudi Arabia passed strict new censorship laws regarding social media. According to the public prosecutor's office, "Producing and distributing content that ridicules, mocks, provokes, and disturbs public order, religious values and public morals through social media will be considered a cybercrime," punishable by up to five years in prison and a fine of three million riyals (nearly $800,000).[21]

When protests swept through Kiev in January 2014, protesters received an identical message on their phones: "Dear subscriber, you are registered as a participant in a mass riot." The Ukrainian government had just passed a law that would send protesters to jail for fifteen years if they were found to be participants in a "mass riot." Meanwhile, in Nghe An Province in Vietnam, Hoang Duc Binh, an environmental activist, was sentenced to fourteen years in jail for livestreaming a video of fishermen demonstrating against a Taiwanese steel company that had spilled toxins in the ocean.

Our phones are also a geopolitical tool. US intelligence, through USAID, created a "Cuban Twitter," a social networking and microblogging site called ZunZuneo, Cuban slang for a hummingbird's call. They hoped to cultivate a sizable user base and then use the network to foment political dissent among island youths, possibly sparking a "Cuban Spring."[22] Other examples of smartphone geopolitics are much more sinister. The data collected by the CIA's Mobile Devices Branch and the NSA facilitates the US government's drone warfare program. "The Drone Papers," an in-depth investigation by journalists at *The Intercept*, details how the CIA and the US military's Joint Special Operations Command have developed a "new global architecture of assassination." Suspected terrorists are tracked primarily using "signals intelligence" and then shot down with a missile from a Predator or a Reaper drone. Operation Haymaker, a special-ops campaign in northeastern

Afghanistan, provides a sense of both how unreliable signals intelligence is and, more broadly, the "collateral damage" of war in the smartphone age: "During one four-and-a half month period of the operation, nearly 90 percent of the people killed in airstrikes were not the intended targets."[23]

As Edward Snowden says, surveillance technologies are "limited to war zones at first," but eventually "surveillance technology has a tendency to follow us home."[24] Technologies developed for military use are used by American police departments and intelligence services to track American civilians. One such technology is "stingrays" (or IMSI catchers)—portable devices that mimic cell-phone towers, tricking any nearby phones into connecting with them, thus revealing their location and identifying information. Police routinely use these devices to pinpoint the location of suspects' phones and thus the suspects themselves.

Stringrays are just one example of how digital monitoring and individualized harassment have become a central part of how US law enforcement polices dissent. The Northern California branch of the American Civil Liberties Union released a report in 2016 detailing how a social media intelligence platform called Geofeedia sold real-time, location-based data about social activists to over five hundred law enforcement and public safety clients. Geofeedia, which has received funding from In-Q-Tel, the CIA's venture capital firm, had "developer access" to Instagram's, Facebook's, and Twitter's data feeds, which gave it the ability to analyze streams of public user posts in real time, including the option to search mentions for specific topics such as hashtags, events, and places.[25] The Baltimore Police Department used Geofeedia to monitor and arrest protesters in the unrest following the death of Freddie Gray, a young Baltimore man whose spinal cord had been severed nearly in half following a "rough ride" in the back of a police van.[26]

After the ACLU report the tech platforms rescinded Geofeedia's access to this data, but as Kalev Leetaru, a senior fellow at George Washington University's Center for Cyber and Homeland Security, argues, cutting off Geofeedia's access means little. Law enforcement, the military, and intelligence agencies have their pick of tech companies offering the exact same product.[27]

Some police departments skip the middleman and do the social media surveillance themselves. A lawsuit brought by the ACLU of Tennessee forced the city of Memphis to hand over documents revealing how local police had created fake Facebook profiles of community activists to give law enforcement access to private posts and to join private groups. This political monitoring comes as no surprise to organizers such as Keedran Franklin, a local activist in Memphis around issues of police brutality. Franklin says police in unmarked cars regularly stake out his office and when Franklin participates in local demonstrations he's often the first person arrested.[28]

American law enforcement and intelligence agencies have a long and storied history of surveilling activists—the FBI tracked Martin Luther King Jr. for decades—and infiltrating antiwar, civil rights, environmental justice, feminist, and labor organizations. Surveillance is a permanent and core element of policing. Today we're witnessing a new chapter in the evolution of policing, both in the United States and around the world. As politics becomes personal governments around the world are cracking down on social media and messaging tools. In dozens of countries people have been put in jail for things they said or posted on social media, even for "liking" materials that were deemed offensive by government officials or for not denouncing messages they'd received.[29]

Governments are working hard to chill dissent, but they face an uphill battle. Political unrest is growing. Since 2010 movements calling for political change have sprung up around the world. Uprisings in Tunisia in 2010 were followed by revolt in Libya, Egypt, Yemen, Syria, and Bahrain. Although these movements were lumped under the umbrella of "Arab Spring," dozens of African countries have seen protests over the past decade. Nigeria saw one of its largest demonstration ever in 2011. Meanwhile in Greece, protesters put a new government in power in 2015. In Spain a neighborhood group called Podemos emerged calling for an "ethical revolution," and Brazilian president Dilma Rousseff was ousted in a whirlwind impeachment in 2016. The same year saw an unprecedented shift in political sentiments in the United States, while Britain shocked the world with a referendum vote to leave the European Union.

Is this political upsurge the result of our new way of doing politics? Are we seeing what happens when politics become personalized—when people have the means to make politics and connect in new ways? This framing is certainly a common way of explaining recent dissent. The Egyptian protesters who gathered in Tahrir Square were constantly asked by journalists to explain how technology and social media were driving their actions; their momentous overthrow of Mubarak was dubbed "the Facebook Revolution."

Smartphones and social media play an important role in recent political unrest, but we risk losing sight of much bigger shifts if we misunderstand how the machine in our pocket is shaping our world. The birth of the iPhone coincided with the biggest economic crisis since the Great Depression, the global financial crisis that began in 2007 and progressed to a full-blown meltdown by the fall of 2008. We can't understand the present political moment without connecting our hand machines to the unsteady terrain of our for-profit system.

In the years following the dot.com crash in 2000, Wall Street players—bankers, brokers, traders, hedge fund managers—were flush with ready access to cash in the low-interest-rate environment and looking around for a new game at the casino. They found one in the US housing market, perfecting a speculative model based on bundling and slicing home mortgage loans, and transforming them into profitable financial products that could be bought, sold, traded, hedged, insured—you get the picture. Pretty soon everyone wanted in on this profitable new game, from towns in Iceland to teachers from California to Saudi billionaires.

The problem was that despite the assurances of the ratings agencies the securities weren't as secure as everyone believed. Their value rested on a shaky pyramid of rising home prices, easy credit, and ordinary people's ability to pay off dodgy and often predatory mortgages. The conditions that fueled the bubble started to fade by 2005, and by 2007 the party was over. When Lehman Brothers declared bankruptcy it set off a global panic that brought down some of the biggest financial institutions in the world. The bleeding was stanched only when the US Treasury together with the Federal Reserve organized a multi-trillion-dollar bailout to calm global financial markets.

In hindsight it's easy to vilify Wall Street. But the wolves were acting rationally in an environment in which the financial sector had been given freer and freer reign to do what it liked, in an economy that had become increasingly dependent on profits derived from machinations in global stock markets. This state of affairs didn't change with the crisis. Once again the federal government bailed out Wall Street and its executives, as it had bailed out banks and elites (in large part through its global institutions) during dozens of financial crises in countries around the world in the previous two decades.

Main Street didn't get a bailout however. When the US economy nosedived ordinary people lost big time. Jobs, and retirement savings that had been squirreled away for years, disappeared overnight. Millions of families lost their homes. Those already on the fringes suffered the most. Half the wealth of African American households—many of which had been specifically targeted for subprime, often fraudulent, loans—evaporated. As the economy "recovered," most visibly in renewed profits for big corporations, many regained their footing. But something had shifted. The size and magnitude of the crisis, and the losses that regular people both in the United States and around the world sustained, coupled with the golden parachutes given to banks and bankers, was too much to bear. People were angry.

Faith that capitalism could provide a good life for working families plummeted in the United States after the 2008 financial meltdown. "Neoliberalism"—a model of capitalism emphasizing a reduced state role in regulating markets and providing services, and "free market" competition between countries, companies, and workers as the best way to achieve growth and efficiency—lost its legitimacy.[30]

In the three decades leading up to the 2008 crisis, advocates of neoliberalism promised to raise all boats through privatizing public services and institutions, shrinking the social safety net, growing the "knowledge economy," bolstering globalization and free trade, deregulating financial markets, and maximizing shareholder value. Democrats and Republicans reassured skeptics that any pain resulting from these policy objectives would be temporary adjustments as the country shifted gears toward developing a high-tech, skilled workforce guided by efficient capital markets and the carrot and stick of "world-class" competition, which would yield good new jobs and economic growth.

Critics pointed to job loss, increased precarity and inequality, and the amoral implications of a shareholder value society, but there wasn't much room for dissent in the Clinton and George W. Bush years. The 2008 crisis changed all this. When the financial sector melted down, a space opened up for popular critique—for people to question the rules of the game and to take a hard look at who benefited and who seemed destined to lose.

Elites tried to paper over the disquiet, declaring the recovery, fueled by quantitative easing, on track. But the crisis opened people's eyes to neoliberalism's broken promises. Criticism of the bailout morphed into a broader anger. People no longer kept quiet about the fact that they were financially worse off than their parents' generation; they opened up about their fears that their children would likely be even less secure. People challenged the common sense that globalization is good, pointing out that for most folks it hadn't brought better job opportunities after destroying the old ones. They expressed anger that elites and corporations seemed to get whatever they needed while requests for decent jobs, community services, and a greener economy were painted as unreasonable or impossible. They demanded to know why rhetorical respect for diversity and multiculturalism had failed to defang a murderous police state, and why women still received less money and respect despite graduating from college at higher rates than men.

Angry about these broken promises and with the help of their phones and social media, people have started organizing. When one looks at the recent decade of global unrest in Iceland, Egypt, Greece, Nigeria, Brazil, Turkey, the United States, and so many other places, a connecting thread is visible. Of course, a thread is thin and flexible; these uprisings are rooted in each country's own history of struggle. Nonetheless, we see common demands. We see a big-picture critique of a system that has enabled elites to capture unimaginable wealth and power while ordinary people have struggled to achieve a secure, dignified, and fulfilling life. Moreover, because these movements are digitally connected, they see themselves in each other.

Modern-Day Revolt

Black Lives Matter is an example of a modern-day revolt against capitalism's violent, racist status quo—a digital-analog movement, or as BLM cofounder Patrisse Cullors calls it, "a social media/all out in the streets" organizing effort.[31] The term "digital-analog" is a shorthand that captures the unique characteristic of modern-day movement building, which occurs on both digital and corporeal planes simultaneously. There is no term that perfectly captures this characteristic, so this shorthand indicating a blending of the digital with the analog will have to suffice for now.

BLM is rooted primarily in the long struggle for freedom and justice by African Americans, a struggle that got rebooted and re-envisioned after the shooting of Trayvon Martin in 2012. The murder brought to the surface deep frustrations over the fact that America finally had a Black president, yet Black lives remained insecure and expendable. But BLM is also a movement that reflects how politics have become more personal and how our always-on digital connection shapes movement building.

In recounting the story of how BLM began, Alicia Garza recalls how she was out with her husband and a couple of organizer friends at a cocktail bar, but keeping a close eye on her phone. Garza, like many others, was waiting for the jury to reach a verdict on whether George Zimmerman would go to jail for killing Trayvon. The bar went silent as the decision popped up in people's Facebook news feeds: Zimmerman would walk. The news was devastating for those who had doggedly demanded justice for the seventeen-year-old boy and his family.[32] Garza said she saw so many messages of resignation on social media that she decided to do something different: she wrote a "love letter" to Black people that ended with, "Black people. I love you. I love us. Our lives matter." And she posted it to Facebook.[33]

Patrisse Cullors, Garza's longtime friend and a fellow organizer, shortened Garza's heartfelt message to the hashtag #blacklivesmatter and began adding it to tweets and posts to encapsulate the broader message that society wasn't respecting Black Americans' right to life and dignity and that it was up to ordinary people to fight for change. Cullors, Garza, and Opal Tometi, executive director for Black Alliance for Just Immigration, an immigrant rights group in New York City, saw the hashtag as a tool to build awareness and solidarity.[34]

The hashtag drifted around on social media until Michael J. Brown Jr. was gunned down by a police office in Ferguson, Missouri, in the summer of 2014. Brown's body lay in the street for hours as police prevented even his parents from going to him; Brown's mother, Lezley McSpadden, had to identify her son's body from a video someone had captured on their phone. It was a tipping point for Black Ferguson residents, who had long been targeted for abuse by city officials and law enforcement. A federal investigation later revealed to the rest of the nation that this abuse included a venal revenue-generating strategy in which Ferguson police filled city coffers by harassing and arresting residents, overwhelmingly Black residents, through traffic stops, parking violations, and arrest warrants.[35]

As millions watched the explosion of anger, Garza, Cullors, and Tometi used the #blacklivesmatter hashtag as a rallying call for a "Freedom Ride" to Ferguson. Hundreds of activists, organizers, and journalists from around the country answered the call, joining the Ferguson residents in the city's fraught streets to demand something better. Photos of those hot and heated weeks show protesters, most with their phones out, standing face to face with a police force equipped for military combat. Movement messages were shared on social media while individuals traded battle stories through their own feeds. On a more practical level, smartphones provided a crucial logistical tool for Ferguson activists, allowing them to keep tabs on the police, coordinate protest strategy, and organize support for those who had been arrested.[36]

Outside Ferguson, people around the country and the world were also glued to their phones, using their own networks and voices to share their own interpretation of the drama unfolding. Indeed, the protests that emerged in response to Michael Brown's murder show the simultaneously decentralized and collective nature of smartphone politics. They are collective in the sense that they generate, through users' digital networks, a shared feeling, a swirling mass of sentiment produced by thousands of tweets and posts and shares that individuals actively generate and share. There is a sense that we are also, or can be, a part of a movement, and that the things happening in our own lives and communities are part of this movement. This feeling was present in past social movements in the United States, to be sure. The Depression-era sit-down strikes, the civil rights movement, the feminist movement, the antiwar and environmental movements, were all movements that drew people into a collective vision. Today we're witnessing the evolution of movement building. Our phones connect us to movements that feel like a living thing, changing and evolving by the second.

At the same time, BLM exemplifies the decentralized nature of the movements that have emerged amid the broader legitimacy crisis of neoliberal capitalism. The #blacklivesmatter concept moved beyond Garza, Cullors, and Tometi's control almost instantaneously. People interpreted and used the sentiment for their own political purposes, sometimes directly at odds with the women's original vision—the #alllivesmatter iteration, for example. Decentralization is also apparent in the absence of a clear spokesperson for the Ferguson unrest. Indeed, when the Reverend Al Sharpton showed up in Ferguson, as he has in other towns and cities when police have murdered Black people, the young organizers who had been in the streets for days were unimpressed with his calls for patience and his insistence on keeping the focus of the protest tightly on Brown's death.[37]

Young people increasingly saw the police killings as part of a bigger systemic problem connected to broader structures of poverty and racism and rooted not just in a racist system of law enforcement but also in deeply unequal systems of education, employment, health care, finance, and political representation. As African American studies professor Keeanga-Yamahtta Taylor contends, "The political action of young Blacks is not happening in a vacuum; it is part of the same radicalization that gave rise to the Occupy movement."[38] It is part of a broader revolt whose dramatic unfolding plays out on small screens and city streets simultaneously.

Like the young people in the streets of Ferguson, the young people who decided to camp out in Zucotti Park in September 2011, sparking the Occupy Wall Street movement, were angry. They weren't angry about police brutality—most were white college students who had little experience with hostile police officers. They were angry about capitalism and their shattered dreams. Many had finished school with a mountain of student debt and no good job prospects; many had to move back home with their parents in a society that blamed them for not choosing a better major or attending a cheaper school or working harder.

The young protesters in New York quickly got acquainted with the police, however, when the NYPD corralled them and doused them with pepper spray. Someone caught the scene on their phone and posted it to social media, slowing down the video to show an officer pepper-spraying a young woman in the face from close range, causing her to double over in pain. What had been a quiet gathering in the park quickly grew as the video went viral; people streamed into the park to show their support and join the encampment. Within weeks similar encampments spread around the US and even around the world, with participants at each site demanding justice and security for ordinary people.

Social movements are rarely islands. They connect through ideas, people, and organizations across time and space. This is true more than ever in the digital age. Occupy failed in the sense that it didn't bring about any change on Wall Street and its encampments collapsed, particularly after police raided Zucotti Park in November 2011. But the systemic critique of the status quo that Occupy inspired, and the digital-analog model it fostered, lived on and blossomed. Bernie Sanders's 2016 presidential run, the Dakota Access Pipeline Protests, and the more recent Sunrise Movement of young people united against global warming are three recent embodiments of a growing progressive upsurge.

Like BLM and Occupy these political constellations seamlessly blend digital and analog actions and organizing; their simultaneously on-the-ground and in-the-cloud existence are written into their DNA. Bernie Sanders crisscrossed the country making speeches; his supporters spent countless hours making phone calls and spreading the word. To build support to stop an oil pipeline from being built under their water supply and through sacred burial grounds, members of the Standing Rock Sioux Tribe in Standing Rock, North Dakota, established a protest camp that thousands of people visited between April 2016 and February 2017, organized a 2000-mile "spiritual run" along with teach-ins and rallies. Sunrise Movement members, in a quest to raise awareness and support for a Green New Deal, went to Washington in late 2018. They sang songs and engaged in peaceful civil disobedience outside the offices of Democratic lawmakers, and fifty-one were arrested.

At the same time, these movements were rooted in the digital. In 2015, 76 percent of Americans had never heard of Sanders or had no opinion of him; by October 2018, more than half of Americans had a positive opinion of him, largely because of social media.[39] Millions followed the Dakota Pipeline protests, a movement started by LaDonna Brave Bull Allard, a Sioux elder, after two attention-grabbing videos were posted online: the well-known progressive author Naomi Klein's interview of Tokata Iron Eyes, a cofounder of ReZpect Our Water, and *Democracy Now!* cohost Amy Goodman's footage of protesters being pepper-sprayed and bitten by attack dogs at the protest camp.[40]

The digital nature of these movements also mitigated traditional constraints on organizing, such as age and resources. Bernie Sanders was the first candidate in American history to run a serious presidential campaign with zero contributions from big business. One enthusiastic supporter even used her Tinder account to ask men to text a donation to Bernie. Anna Lee Rain Yellowhammer was just thirteen and Tokata Iron Eyes twelve, yet they were key organizers in the Dakota Pipeline movement. Sunrise members are high-schoolers and college students scattered across the country, yet the organization has succeeded in starting a conversation about the sustainability of our for-profit system that the Democratic Party can't ignore.

The digital-analog quality of these movements also ties them to each other and to millions of ordinary people. These ties are visible in the movements' networks of support and inspiration. Black Lives Matter and Bernie Sanders were both vocal supporters of the Standing Rock protests. Young members of Sunrise cite Occupy, BLM, March for Our Lives, and United We Dream, a youth-led immigration justice organization, as sources of inspiration.[41] Congresswoman Alexandria Ocasio-Cortez of New York was inspired to run for elected office by a visit to the Standing Rock camp and is a champion of Sunrise. She worked for the Bernie Sanders campaign and is a member of Democratic Socialists of America, an organization that went from a sleepy five thousand members in early 2016 to more than sixty thousand today.[42]

The shared networks of people and supporters that undergird these movements are also increasingly connected by a critique of capitalism. The landscape of ideas has shifted since the 2008 crisis, evolving in concert with these movements toward a broader critique that links economic, social, political, and ecological concerns organically. This is partly a result of on-the-ground organizing, and partly the result of new ideas and sources of information being distributed and cultivated online through social media.

Sanders's popularity was the result of a deep sense of distrust of the establishment and an impatience with the rhetoric of status quo elites from both parties who campaigned on a platform of "Everything is fine. Chill."[43] But Sanders wasn't the only one who benefited from this distrust. Donald Trump also appealed to people who were on the losing end of neoliberalism, and unlike virtually any Republicans before him, Trump also relied heavily on small donors. But unlike Sanders, who was coalescing a radically progressive program, Trump was pulling together threads from the right. While Sanders talked about single-payer health care and free higher education Trump fed his networks stories of Mexican rapists and nostalgia for a time when America was "great."

Trumpism, which political scientist Cas Mudde describes as a "radical right combination of authoritarianism, nativism and populism," can be traced back to a response to the 2008 crisis, but one very different from the response on the left.[44] After the crisis anger over debt, precarity, the bailout, job loss, and alienation was also articulated through right-wing voices such as Rick Santelli, now an editor for CNBC. In his famous 2009 rant from the trading floor of the Chicago stock exchange, Santelli blamed the crisis on crony capitalism—the government's symbiotic, parasitic relationships with businesspeople too lazy or unimaginative to compete successfully in the marketplace—and losers looking for a handout.[45] Instead of trying to buy our way out into prosperity, Santelli shouted, America needed to "reward people who carry the water, instead of drink the water." Santelli implored the country to return to its bootstrap roots.[46] On April 15, 2009, tax day, "Tea Party patriots" energized by Santelli's speech held protests around the country demanding a return to fiscal conservatism.

The Tea Party, like Occupy, spawned a cluster of new right-wing figures and movements that have their own systemic critique blending cultural, political, and economic factors. Steve Bannon, Trump's campaign manager and briefly the chief strategist for the White House, calls for economic nationalism, infrastructure programs, and closing the borders to immigration to prevent the collapse of both capitalism and the "Judeo-Christian West." Bannon's worldview echoes that of the paleoconservative William Lind, who warns of a "fourth-generation war" in which, instead of being invaded by armies, the United States will be invaded by immigrants, sucked under by the "poisonous ideology of multiculturalism."[47]

White supremacy has seen a resurgence, evidenced by the popularity of Richard Spencer, the president of the National Policy Institute, a white

supremacist think tank. Calls for men to be men again have also seen renewed favor. Gavin McInnes, the cofounder of Vice Media and onetime
leader of the Proud Boys movement, characterizes his ethos as "libertarian
politics, father-knows-best gender roles, closed borders, Islamophobia, and
'western chauvinism.'"[48]

White supremacy, xenophobia, sexism—these ideas are as American as
apple pie. But right-wing ideals have seen renewed fervor in the decade since
the financial crisis: They address the fears and frustrations of a different
segment of the American population angry over declining living standards,
and alienation. They combine old villains (people of color, immigrants, socialists, feminists) with new ones (Muslims, fake news, Silicon Valley elites)
to weave a modern narrative about the destruction of American jobs and
culture, the weakening of our country's virility and fortitude.

The rising right worldview is orthogonal to that of the left, but its
movement-building model since the financial crisis has some similar digital-
analog features. Twitter, Facebook, YouTube, 4Chan image boards such as
/pol/, subreddits, private chatrooms, news sites such as Breitbart, Drudge
Report, and Gab (a social network with eight hundred thousand users) have
fostered a constellation of right-wing voices fused together through memes,
wink-and-nudge emojis, jokes, and promotional videos. The Rise Above
Movement (RAM), an "alt-right fight club" that originated in California,
posts promotional videos for devotees and potential recruits featuring mon-
tages of RAM members practicing mixed martial arts while wearing skull
masks. ProPublica linked the murder of a young gay Jewish man in California to members of a white supremacist group who interacted in private chat
rooms. The media scholar Jonathan Albright, after examining the thickets
of right-wing messaging networks that have developed in recent years, says,
"Rightwing news is everywhere."[49]

White supremacists, neo-Nazis, and other members of the far right have
long cultivated an online presence. But in recent years right-wing movements have been moving off the web and into the street. In August 2017
James Alex Fields Jr. killed a counterprotester, Heather Heyer, and injured
several others with his car shortly after the Unite the Right white supremacist rally in Charlottesville, Virginia. The attack and the rally are part of a re-
kindling of white supremacy that began with the election of Barack Obama
and has seen a significant uptick since then. Nearly a hundred people were
killed in terror attacks in 2017; officials say 60 percent of those deaths were
caused by people who ascribe to right-wing ideologies.[50] The Rise Above

Movement made headlines in 2017 for sending its members to beat up antifascist protesters at rallies in California and to the white supremacist rally in Charlottesville. McInnes's Proud Boys, some of whom also attended the Charlottesville rally, have beaten up protesters in New York City and Portland, Oregon. Eight members face assault and rioting charges in New York City.

This surge of angry white men—or boys, if we follow McInnes's nomenclature—demanding the restoration of their right to absolute power has been matched by a surge of angry women. The election of a man who boasted about the pussy-grabbing rights conferred to him as a result of his power and wealth threw a fat log on a growing fire of feminist fury in the United States and around the world.

The feminist movement never goes away, but it waxes and wanes. Since 2010, feminism, like many other social movements, has been on an upswing. Women have begun using the "F" word again. In a 2013 *Rolling Stone* interview, Beyoncé called herself a "modern-day feminist" and Sheryl Sandberg published *Lean In*, a runaway hit in which she implored women to take off their tiaras and take the corner office.[51] Sandberg's vision of feminism—that women are on the cusp of achieving equality if only they'd "put their foot on the gas rather than the brake" in their work lives—became a dominant frame for women's liberation. If each woman strived to take a position of power, in her workplace and community, collectively women could use this power to reshape practices, laws, and norms for the benefit of all women.

Many women believed that Hillary Clinton, who seemed anointed to become the first woman president of the most powerful country in the world in 2016, would have actualized this vision. But Clinton lost. Her defeat was a major disappointment for millions of women. It was also a turning point, marking the high water mark of neoliberalism feminism—the idea that we can achieve the goals of feminism by individual women striving to get ahead in the system. The Women's March following Trump's inauguration, one of the largest protest marches in American history, was as much a funeral march for this blueprint for feminism as it was a protest against the new man in the White House.

Women *are* taking power. As Keeanga-Yamahtta Taylor observes, "The face of the Black Lives Matter movement is largely queer and female."[52] The same could be said for progressive social movements more broadly. In place of neoliberal feminism a new, collective politics of feminism is emerging and it's intimately tied with our new way of doing politics. Women are using their networks, and building new networks, to amplify their voices, their anger, and their pain. Instead of demanding the corner office, they're demanding that the guy in the corner office who rubbed his penis against them, or tried to trade sex for a promotion, be kicked to the curb. They're turning "whisper networks" into scream, shout, and yell networks.

This collective rage has been building for a decade: Slutwalks in Toronto, the March of the Margaridas in Brasília, the "Feminist Five" in China are but a few examples. But the personalization of politics—the interspersing of politics into the nooks and crannies of our daily lives through our constant digital connection—pushed this rage into an unexpected, though not entirely surprising, direction in October 2017. A story about Harvey Weinstein and his ugly, decades-long mistreatment and assault of women was published in the *New York Times*.[53] The flapping-bathrobe narrative was a familiar one, but it landed in a new landscape, one in which women had had enough of abusive men in power. Women turned their personal digital connections into a collective "Enough!" in what has become the #MeToo movement.

Tarana Burke created the #MeToo hashtag a decade ago. Other woman have also used their pocket computers to spread female solidarity: Jessica Knoll's #WhatIKnow, and Kelly Oxford's #NotOkay hashtags both called out sexual assault and encouraged women to own their pain and connect with other women who had experienced assault. But in the years since Burke coined #MeToo, our relationship with mobile social media has deepened. We share more and more. The sharing that we do on social media—jokes, memes, bad-date stories, kitchen makeover pinspiration—has extended to something deeper, something political. Our digital connection has created the idea that there are people who will listen.

They are listening. Celebrities, politicians, and men who've used their power to abuse women are being called out and publicly shamed in an unprecedented way. The reckoning is unplanned and leaderless. It abides by no set guidelines and makes both men and women deeply uncomfortable, a discomfort reminiscent of earlier waves of feminism. No one is sure where #MeToo is going, whether it will maintain its momentum or fizzle. Regardless, it exemplifies how the political has become personal.

Finding Our Voice

The personalization of politics afforded by mobile social media has created a new way of doing politics that is central to the movements that have emerged in the past decade. But like all politics, it's messy. One of the messiest elements of the emergent digital-analog model is how explosive politics has become. Love turns to rage in an instant. Viral whiplash surrounded Keaton Jones, the kid whose mom posted a video of Keaton tearfully describing being bullied at school. One moment the boy was awash in public sympathy and affection, the next he was collectively scorned when a photo of his mother posing in front of a Confederate flag was posted online.

Writing in the wake of public furor over whether the children's show *Peppa Pig* created unreal expectations of Britain's National Health Service, Ashley "Dotty" Charles, a UK rapper and BBC announcer, warned against the increasing prominence of outrage politics. As a Black, gay woman Charles said she had learned to become selective in her anger: "If we are all outraged all of the time, then outrage simply becomes the default setting. By shouting about everything, we are creating a deafening silence where outrage is without consequence."[54]

Moreover, with so much of our smartphone politics defined by digital engagement with people and networks it becomes difficult to parse genuine political expression, debate, and movement building from "virtue signaling" and narcissistic self-promotion. The personalization of politics has made politics extremely performative. More and more, political ideas and movements are associated with YouTube and Twitter personalities. With so many people and personalities on the web, and a growing obsession with "likes" and followers, many choose to differentiate themselves by appearing to be the most righteous, whatever their political persuasion, engaging in flame wars of little political value. It's the next iteration of politics as spectacle.

We can't blame narcissism for everything, however, tempting as it may be. Some of the mayhem is caused or enhanced by digital profit-generating strategies and algorithmic mischief. For example, bots are automated programs that run on the internet and pretend to be humans and, sometimes, are designed to push people's political buttons. "Jenna Abrams," a memorable bot who managed to garner seventy thousand followers on Twitter, was created by a troll farm, the Internet Research Agency, in St. Petersburg, Russia. The Trump-loving bot's calls for resegregation and efforts to delegitimize "manspreading" had thousands of people taking time out of their day to argue with "her" online.[55]

The tech titans also want to keep us on their sites, so they use their algorithms to feed us what they think we want to see. Cass R. Sunstein, a Harvard Law School professor, worries about the polarizing impact this insulation will have on the body politic. He says, "Social media makes it easier for people to surround themselves (virtually) with the opinions of like-minded others and insulate themselves from competing views." Likening social media to a disease vector, Sunstein surmises that it is "potentially dangerous for democracy and social peace."[56] In these days of filter bubbles and algorithmically generated search results it takes effort to seek out opposing political views. Our social media feeds largely show us political content that we already "like." If we don't hear other people's perspective, can we agree on a collective political project?

Others say much of the behavior associated with modern social movement organizing is not real politics. They call it "slacktivism," a term coined by the technology scholar Evgeny Morozov to describe how we think we're changing the world with our online engagement but we're actually doing nothing, or worse than nothing because if we weren't fooled into thinking that our smartphone politics mattered maybe we'd be out in the streets, or in our communities, doing something real.[57] A conversation between comedians Jerry Seinfeld and Trevor Noah on the Netflix show *Comedians in Cars Getting Coffee* lays out this popular position:

SEINFELD: Doesn't it seem like we're striving to take the entire life experience and have it in our underwear? You can socialize, do work, get entertained, and get information, all in your underwear.

NOAH: People are now able to protest in their underwear. And that almost defies what protesting should be about. The whole point to a protest is to get up out of your bed, put your clothes on, walk out into the cold, and say, "I stand for this, I march for this." But now you don't really have to have that conviction, because you're on your couch, in your underwear. . . . Punch in a few characters and go "Yeah, yeah, I fought for the cause." No you didn't.[58]

Are the digital-analog movements that have emerged just fake politics? We can point to many failures of the recent digital-analog upsurge. The #NeverAgain students failed to elect a pro–gun control governor in Florida in 2018. Bernie Sanders failed to win the Democratic Party nomination in 2016. Trump reversed the halt on pipeline construction in North Dakota.

Black Lives Matter has yet to achieve measurable police reform and few police officers have gone to jail for murdering Black people. The right, too, has
had its own failures. The jubilant tiki-torch-wielding fascists who marched
in Charlottesville have been driven back into their holes as Richard Spencer,
Alex Jones, Gavin McInnes, and numerous other representatives of the right
have been "deplatformed" by the major social media companies.

Criticisms that paint the digital-analog political model that has emerged
in the past decade as primarily an artifact of filter bubbles, virtue signaling,
and slacktivism have some validity. The power that the tech titans have to
shape and censor our political reality, particularly because these platforms
have become a primary source of news, is deeply concerning. At the same
time, both companies and states are trying to control how we use our phones
to do politics. Censorship, surveillance, and the contraction of the web as
a free place to explore ideas are serious problems that require a concerted,
organized response.

But we should be wary of technologically determinist explanations of the
current political landscape. American society is deeply polarized because it
is in crisis—neoliberal capitalism has lost its legitimacy. As political scientist
Nancy Fraser says, ordinary folks have "lost confidence in the bona fides of
the elites" and are "searching for new ideologies, organizations, and leadership."[59] This legitimacy crisis is not an artifact of our filtered social media
streams. People are using their phones to find their voice in capitalism—to
challenge the status quo and explore political alternatives. Polarization may
be fostered by the architecture of our phones and platforms, but this doesn't
mean the political unrest that characterizes the present moment is algorithmically generated.

It is also true that people perform politics for attention, and that this
theatrical propensity has evolved and intensified in the transition to smartphone politics. But instead of chalking up the "failures"—particularly of the
progressive movements we've focused on in this chapter—to the limitations
of slacktivism, it is useful to situate the failures in a broader context. Indeed, the presence or prevalence of slacktivism is something that needs to
be explained.

Ultimately, the problems the left is tackling—racism, sexism, gun violence, climate change, inequality, poverty—are not problems that can be
solved on Twitter. But they're also problems that haven't been solved in the
voting booth, or in the street. It's not even clear what the way forward is,

in some cases. The social movement theorists Adam Branch and Zachariah Mampilly note that the "lack of alternatives to the current order obscures the relation between political grievances and objectives. What exactly needs to change in order to bring about desired transformations in people's lives is often unclear."[60] But this isn't a reason to despair or to decide that the emergent digital-analog model of doing politics is meaningless or fake politics.

We know that successful social movements require on-the-ground institutions and networks of real people we can speak to on the phone or better yet, get in a room together. Alexandria Ocasio-Cortez, the youngest woman ever to be elected to Congress, won through "shoe-leather organizing" *and* social media. AOC and her team made over 170,000 phone calls. They went door to door for months, getting phone and contact information, and distributed literature all over her district, stopping to talk with people and if possible bring them into the campaign. Social media and the digital half of the emergent digital-analog political model was only one piece of AOC's campaign. But it was an important piece.

In the sixties and seventies, numerous mass movements demanded a response from elites: farm workers, organized labor, feminists, the antiwar movement, the environmental movement, the civil rights movement. Despite being powerful and well-organized, they were crushed, delegitimized, and defanged as a result of both a shifting social, economic, political landscape in a decade of crisis and their inability to resolve deep internal issues about how to build an inclusive working-class movement. Organizers and activists building political power through their smartphones today are grappling with both the divides of the present and the failures and mistakes of past social movements. They're developing new movements and a new digital-analog organizing model that, taken together, have the potential to foster a political coalition that is both radical and inclusive.

To be sure, the power of the left to achieve gains for working people is severely diminished relative to the sixties and seventies. In neoliberal capitalism centrism ruled and Utopia became taboo. Desire for something better than capitalism was considered a sign of hysteria or unconscious authoritarian impulses.[61] Nonetheless, ten years after the financial crisis the left is finally emerging from its "deep-rooted economic fatalism."[62] People are starting to question what went unquestioned for the past four decades: that markets are always right and good, that unions and social movements are a hindrance and a burden, that the government should exist primarily to create smooth pathways of wealth accumulation for business and elites. Progressives are uncertain of the horizon they seek—they are in a phase of consciousness-raising, building from scratch. But meaningful social change, for the first time in a long time, seems possible. As politics become personal people are beginning to feel like their voice matters. A new legitimating framework is crystallizing. The question is: Who will determine the horizon of ideas in our smartphone society?

Chapter 5

New Spirit

Mamoudou Gassama was on his way to watch the Champion's League Final with a friend when he walked into a big commotion in Paris's Eighteenth arrondissement. A panicked crowd had gathered below a toddler dangling from a balcony high above the street. Gassama sprang into action, climbing up the side of the building and saving the child. Bystanders filmed the feat on their phones and put it on social media. Overnight, Gassama, a young Malian refugee, became a French national hero—a real life Spider-Man.[1]

Our phones are storytelling machines. They bring us not only stories of heroes, like Gassama, but also gossip, news, and cautionary tales about friends and family, pop stars, and professional athletes. Equally important, they let us tell stories about ourselves. Through text, image, and video we craft personal narratives that make us appear beautiful, clever, and responsible, stories that project competence and empathy. Other times we stray from Instagram filters to tell darker stories that express heartbreak and anger, disgust and disappointment.

It's not surprising that we use our hand machines for telling stories. Stories are a big part of what makes us human. We express our beliefs and values, find friends and lovers, and teach our children through our stories. In fact, society depends on stories—both small stories about our personal hopes and dreams, trials and tribulations, and bigger stories about human nature, good and evil, and the future horizon. The stories societies tell about where they've been, where they're going, and what they value are central, and unique, to each generation and each society. They shape the contours of our collective consciousness and provide a backdrop of meaning and mooring for our personal biographies.[2]

Modern society is particularly in need of meaning and mooring. Although we don't often articulate it, we live in a capitalist country where profit is the driving force. Yet most people aren't particularly motivated in their lives by profit. We care most about our families, friends, communities, and ideals (God, country, sports). These contradictory drives make strange bedfellows: humans are social, mutualistic beings, yet we're locked into a system, wage earners and capitalists alike, designed to perpetually accumulate more and more profit, rather than satisfy human needs or provide for the common good.[3]

Why do we submit to such a contradictory way of organizing society? To be sure, the poorest and most desperate among us have no choice; or, rather, the choice is submit or starve. But compulsion is insufficient for generating sustained profits, growth, and innovation. The amazing breakthroughs of the capitalist age cannot simply be written off as products of coercion and duress. For capitalism to exist, ordinary people must find meaning in our for-profit society; they must willingly direct their creativity, energy, and passion toward their work; they must actively, or at least passively, believe that societal structures are capable of meeting their need for justice and security.

We don't draw a sense of meaning and justice from the drives of capitalism itself; there's nothing intrinsic to profit making that provides these things. Indeed, the centrality of profit making in the organization of our society is usually left unsaid; capitalism is experienced more as an anonymous force, not as something we agree to or vote upon. To attain people's devotion or at least moderately cheerful submission, capitalism repurposes cultural ideas and values from *outside* the circuits of profit making—honor, duty, family, creativity, perseverance, responsibility—and transforms them into a legitimating framework that the sociologists Luc Boltanski and Eve Chiapello (drawing from the German sociologist Max Weber) call "the spirit of capitalism." The spirit of capitalism is the set of cultural beliefs associated with capitalism that helps to justify and legitimate it, and, equally important, "to sustain the forms of action and predispositions compatible with it."[4]

As cultural norms and values have changed over time, so too has the spirit of capitalism. As Tim Wu says, "Capitalism is a perfect chameleon; it has no disabling convictions but profit and so can cater to any desire, even those inimical to it."[5] This chameleonesque quality—the periodic emergence of a new set of legitimating principles that facilitate the willing participation of society—accounts for capitalism's remarkable longevity despite periodic bouts of deep crisis. Bursting bubbles, tanking currencies, scuffles between haves and have-nots often make for rough going. When times get really rough ordinary people lose faith in the legitimating framework undergirding the status quo. They start to ask why things must be the way they are, why people can't have something different or better. Capitalism is a historical system and thus is always evolving, but in these moments it must take a leap, making big changes not only to norms, regulations, and processes of redistribution, but also to ideas. Its survival depends on it.

These leaps have occurred numerous times in the past. Older readers will recall the chaotic decade of the 1970s—a decade characterized by stagflation, gas lines, consciousness raising, civil rights marches, and mass strikes. Less remembered is the pervasive sense of dissatisfaction that permeated society.

In the dusk of the postwar boom Americans began to chafe against the limits and restrictions of their workplaces and society more broadly. They grew tired of stifling expectations to become a company man or a good little wife and bemoaned the lack of space for individualism, creativity, and flexibility. Wage and benefit increases weren't enough—people wanted more. Consumers wanted quality and fairness. Women and people of color wanted respect and opportunity. Workers wanted autonomy and space for creativity. These desires, in the midst of a serious economic crisis, created a deep political crisis that ultimately saw Keynesianism thrown into the dustbin of history.

In its place a new set of ideas emerged—a new spirit of capitalism embedded in aspirational stories about flexible workers, nimble firms, and entrepreneurs eager to think outside the box. Knowledge, cutting-edge technology, multiculturalism, and a new global order were emphasized. The state, with its regulations and cronyism, was de-emphasized. Instead of picking winners and losers, the federal government would stop intervening, using its power only to reduce barriers and promote cleansing competition to weed out or wake up those who'd grown smug and lazy. The individual, freed of "big government" could pursue her own interests, while the market, freed of Keynesian fetters, would work its magic creating efficient economic growth and happy, liberated citizens.

The 2008 crisis brought home the hollowness of these neoliberal ideals. When the stock market tanked, bankers got bonuses while regular folks got pink slips. Instead of creating growth and prosperity the "unfettered market" created a giant mess that taxpayers were expected to clean up. This led not only to an upsurge of political organizing but also to a search for new ideas as popular faith in the legitimating framework of neoliberal capitalism evaporated. Margaret Thatcher's mantra that "there is no alternative" to neoliberalism no longer sounded triumphant; it sounded dismal and nihilistic. The "upturned morality," says political scientist Mark Blyth, in which the "naked self-interest of financial market actors was taken to be the most positive virtue because its pursuit led to optimal outcomes despite moral intention" became deeply suspect.[6] People began to openly question why shareholders were so much more important than stakeholders.

Even after the Federal Reserve steadied the financial ship, neoliberalism didn't regain its legitimacy; the Great Recession was followed by the Great Pessimism. An American dream defined by dead-end jobs and debt, by plutocracy rather than meritocracy, wasn't something ordinary people wanted to get behind. Capitalism needed a new legitimating framework—a new spirit. For the past decade this spirit has been in formation and flux. Conflicting voices and stories clamor to be heard and claim the mantle of the future. Not everyone's stories have the power to be heard equally—some people have a lot more power to make sure that their stories are heard. These days Silicon Valley and its high-tech companies are holding the megaphone.

Moonshots and Teddy Bears

Despite its deep roots in the military and ongoing partnership with the American intelligence establishment, Silicon Valley has long projected a wild and free image—a land of tech hippie-cowboy hackers embodying the rugged individualism of "the West" mythologized by Frederick Jackson Turner: "that coarseness and strength combined with acuteness and inquisitiveness; that restless, nervous energy . . . that dominant individualism, . . . that buoyancy and exuberance which comes from freedom."[7]

Ellen Ullman, with the clarifying gaze afforded by a brilliant mind and the experience of being one of the few female software engineers in the Valley in the 1990s, recalls meeting someone who seemed to be the embodiment of Turner's archetype: a "rebel cryptographer" sporting long hair and baggy jeans named Brian: "In appearing to be a genius on a skateboard," Ullman recalls, "he couldn't be playing his part better."[8] Brian cut quite a different figure from the slick haircuts running the show on Wall Street. He embodied a worldview that looked beyond quarterly returns, that saw companies as something more than a stream of assets, saw people as more than a locus of human capital.

Tech giants also seemed to embody this worldview. In Amazon's first letter to its shareholders in 1997 Jeff Bezos declared, "We will continue to

make investment decisions in light of long-term market leadership considerations rather than short-term profitability considerations or short-term Wall
Street reactions." In a 1998 research paper, Larry Page and Sergei Brin, the
Google founders, wrote, "We expect that advertising funded search engines
will be inherently biased towards the advertisers and away from the needs
of consumers." The internet search company instituted a two-class share
structure so investors could profit from but not determine its long-term
strategic plans.

This alternative view took a hit in 2000, when the tech bubble burst
and trillions of dollars in valuations were lost in the span of a few months.
Observers wondered how anyone could ever have believed that Pets.com's
electronic shop window was a viable business model and worried that the
promise of the internet was mostly hype. Maybe the economist Paul Krugman's prediction that the internet's impact on the economy would be no
greater than the fax machine was right.[9] The massive infrastructure build-
out of cell phone towers and fiber-optic cable crisscrossing the country certainly seemed like overkill, a waste of excitement and resources. After the
crash it was back to fundamentals.

But Silicon Valley didn't go bust. It went from Web 1.0 to Web 2.0. The
tech companies that survived grew stronger and more confident, and when
Wall Street banks tanked in 2008 the ideas of Silicon Valley felt like a breath
of fresh air. When the Wall Street masters of the universe, obsessed with
quarterly returns and shareholder value, went hat in hand to the Fed for
a bailout, capitalism suddenly appeared jaundiced and unimaginative. Observers lamented the long-term loss of vision—instead of flying cars we got
derivatives.[10] Capitalist vision wasn't dead however; it had simply moved to
the other side of the country. In an about-face from the ideological imperatives of the 1990s and early 2000s, new companies such as Google, Uber,
Airbnb, and Amazon seemed blithely unconcerned with profits. They told
stories about moon shots, abundance, and living on Mars, and in the midst
of the Great Pessimism we were enraptured. NYU marketing professor,
Scott Galloway, says of the iPhone: it "was a bright light in the darkness that
signaled hope and optimism."[11] It had suddenly become acceptable to think
big again.

Larry Page, whose company's unofficial slogan is "Don't be evil," talks
about using knowledge to fix things and help people. "In every story I read
about Google, it's about us versus some other company or some stupid
thing," he laments. "I don't find that very interesting. We should be building

great things that don't exist. Being negative is not how we make progress."[12] Airbnb cofounder Brian Chesky speaks of the company's "pure" mission to "solve a problem and to help people."[13] Yelp gives its employees teddy bears for being nice.

In the decade after the crash the ideological center of gravity seemed to have shifted. Shervin Pishevar, an "angel investor" and member of the UN Foundation's Global Entrepreneurs Council, articulated the new view: "Where there were once the founding fathers of democracy, there are now startup founders. If Alexander Hamilton were coming from a little island today, would he go to New York City? No, he would be arriving in Silicon Valley."[14]

Societies change. Perceptions and institutions are reorganized; norms and traditions evolve; fads and passions go out of fashion or are incomprehensible to outsiders. Still, in each culture there are also persistent ideals and stories—ideas whose core principles get recycled and repurposed over and over because they resonate with people. A successful spirit of capitalism will draw on these bedrock ideals. The historian Richard Weiss has argued that one bedrock ideal in American culture is "that every child receives, as part of his birthright, the freedom to mold his own life." It is the "belief that all [people] in accordance with certain rules, but exclusively by their own efforts, can make of their lives what they will."[15] The emergent Silicon Valley spirit is, or at least has been for the past decade, so successful because it rebooted this core American ideal, this new expression of the American Dream. It didn't just challenge the morals and drives of the Wall Street titans who personified the neoliberal spirit. It offered something in their place, a new aspirational story that everyone could be a part of—through their phones.

Silicon Valley, we're told, is building an architecture for inclusive growth. Its platforms promise to not only enhance connection but to bring a new world of interpersonal trust and innovation.[16] Each of us can participate easily, seamlessly, through our phones. By creating online profiles, sharing, and connecting with others in new decentralized, peer-to-peer networks the individual can shine and thrive, gaining not only attention and followers but also power and potentially money. These networks of empowered individuals are creating something from nothing—new consumption experiences, new economic activity. We are growing the pie, making new space for individuals who were left out in neoliberalism. As Arun Sundararajan, New York University professor and the author of *The Sharing Economy*, enthuses:

[Thomas] Piketty's "renters" can begin to experience the other side of the coin by making money through investing or owning rather than laboring. . . . People once relegated to laboring for others are assuming new roles and occupying new locations in the established economic equation, transcending from being people who receive wages to people who own capital.[17]

This vision jettisons the false promises of neoliberalism in favor of a new promise: in the digital world being created by Silicon Valley individuals can be their own boss, making capitalism work for them, without relying on the government for a handout or a big corporation for sustenance. A sense of meaning and social justice is restored to the accumulation process. Meritocracy and the dream of entrepreneurship—the republican ideal of free labor, free soil, of escaping the wage relation—are reborn.

Our phones aren't just resurrecting the American dream for users. They're also resurrecting it for companies and investors. Start-ups and unicorns, many of whom have never earned a dime in profit, craft stories of warp-speed growth rooted in network effects and low marginal costs of distribution. They promise to remake markets and reengineer society through mobile tech and killer apps and venture capitalists are enthralled. They could be getting a piece of the next Facebook, the next Slack (a popular workflow platform), the next Uber. Individual investors such as Carl Icahn, venture capital firms such as Andreessen Horowitz, and funds such as Softbank's Vision Fund and the Collaborative Fund have dumped hundreds of billions of dollars into tech start-ups over the past decade. The Silicon Valley spirit is even bigger than the promise of its platforms to create value and success for individuals and firms, however. Silicon Valley, we're told, is ushering in a new age—a digital age of limitless possibility.

The big ideas being cultivated in Silicon Valley have been around for a while. In the early nineties scholars discussed the "New Economy" and the "network society," predicting massive societal shifts due to advances in digital technology and internet connectivity. These ideas can also be traced to much earlier work; scientists in the 1950s believed they were on the cusp of developing a general-purpose artificial intelligence: the ability for a machine to "learn" and in principle perform any intellectual task that a human can. The excitement today stems from the feeling that the promise of these earlier beginnings has finally arrived.

Smartphones, which are "1 million times cheaper, 1000 times more powerful, and about 100,000 times smaller than one computer in MIT in 1965," seem to exemplify this arrival.[18] The explosive growth in connectivity at all scales over the past decade, in combination with advances in data collection and storage, computing power, and machine learning, have convinced many scholars that we're on the precipice of something fundamentally different and new. Two MIT professors, Eric Brynjolfsson and Andrew McAfee, say we've reached the "second machine age." Brynjolfsson and McAfee liken the impact of computers, big data, peer-to-peer networks, and other digital advances in our technological capabilities to the monumental societal impact of the steam engine and electricity:

> However, unlike the steam engine or electricity, second machine age technologies continue to improve at a remarkably rapid exponential pace, replicating their power with digital perfection and creating even more opportunities for combinatorial innovation. . . . The fundamentals are in place for bounty that vastly exceeds anything we've ever seen before.[19]

Chris Anderson, the editor-in-chief of *Wired* magazine, predicted back in 2008 that we'd reached the "end of theory"—that our 24/7 connected lives would generate so much data that soon we wouldn't even need theory anymore: "Correlation supersedes causation, and science can advance even without coherent models, unified theories, or really any mechanistic explanation at all."[20] With advancements in networks and computing we'll generate more and more data, solving problems and fueling breakthroughs in medicine, science, agriculture, policing, and logistics. Self-driving cars and the ability of the Alpha Go Zero program to teach itself how to play the ancient and extremely difficult Chinese game of Go using only the game rules and reinforcement learning are just the beginning of a seismic shift rooted in the power of data.

The motto of Alphabet subsidiary DeepMind encapsulates this vision of the future: "Solve intelligence and use that to solve everything else." Run by Demis Hassabis, a neuroscientist, video game developer, and former child chess prodigy, and a team of about two hundred computer scientists and neuroscientists, the Alphabet subsidiary's researchers have operationalized the idea that intelligence, thought, and perhaps even consciousness are nothing more than a collection of discrete, local processes that can be "solved" with enough computing power and data. Hassabis is one of a growing number of scientists who say the artificial general intelligence we were promised so long ago is finally within reach.

This core belief in the power of technology and data is part of a broader worldview encapsulated in popular Silicon Valley sayings such as "Move fast and break things" and the abbreviated "Ask for forgiveness, not permission." Google never asked permission to photograph the front of everyone's home, link it to a physical address, and put it on the web. Nor did the company ask permission for its Google Earth cars, kitted up with special equipment, to vacuum up unencrypted Wi-Fi traffic (usernames, passwords, emails, photos, videos) as they drove around taking photographs. Incidentally, in that case, Google didn't ask for forgiveness either; it claimed that it had a right to the data, just as someone has the right to listen to a radio station in their car.

In the Silicon Valley worldview, change is almost always good, the more "disruptive" the better, because disruption enables us to break down old barriers and assumptions, an inherently progressive development. Technology is at the center of this worldview because proponents believe that nearly every problem can be solved by education, communication, and technology. Sure, people and countries may still be divided in the New Gilded Age, but these divides are not fundamental. There is a technological fix—we just need to find it.

Some of these guys and gals are such true believers that they have even developed tech-based religious beliefs. Anthony Levandowski founded the first church of artificial intelligence based on a new religion called Way of the Future, a belief system created "to develop and promote the realization of a Godhead based on Artificial Intelligence." Levandowski thinks humans are generating a new god—not a strike-you-down-with-lightning kind of god, but a god nonetheless: "If there is something a billion times smarter than the smartest human, what else are you going to call it?"[21]

Ray Kurzweil, inventor, futurist, and recipient of the 1999 National Medal of Technology and Innovation (granted by the US president), is a proponent of the idea of the Singularity. The Singularity is a hypothesis, shared by prominent scientists and futurists, contending that we're walking down a technological path that ends with an artificial superintelligence so powerful it will someday have the power to reproduce itself and remake human civilization in unfathomable ways. Kurzweil, an optimistic "transhumanist," thinks this future will manifest itself as the merging between humans and computers, allowing human consciousness to live forever. In preparation, Kurzweil has collected every scrap of material possible related to his deceased father so that one day he'll be able to recreate his father's consciousness in the cloud and once more have a chat with him.[22]

There's a catch, however. To get to the future, Silicon Valley thought leaders believe society needs to remain free from the suffocating grasp of the government. Ellen Ullman sees this Ayn Randian sensibility expressed often in the Valley by men who see the government as "anathema, a pit, the muck in which dreams of changing the world will forever sink."[23] In his early days at the helm of Facebook Mark Zuckerberg listed himself as "enemy of the state." Katherine Losse, in her memoir about working at Facebook, recalls an encounter with Zuckerberg in which he asked her to write a series of blog posts about "the way the world was going." He wanted one post to discuss the theme of "companies over countries." When Losse asked for clarification Zuckerberg explained, "It means that the best thing to do now, if you want to change the world, is to start a company. It's the best model for getting things done and bringing your vision to the world."[24] Zuckerberg's sentiment echoes Electronic Frontier Foundation founding member John Perry Barlow's oft-cited manifesto: "Governments of the Industrial World, you weary giants of flesh and steel, I come from Cyberspace, the new home of Mind. On behalf of the future, I ask you of the past to leave us alone. You are not welcome among us. You have no sovereignty where we gather."[25]

Peter Thiel shares Zuckerberg's and Barlow's desire for autonomy and space to build new Utopias. He's bankrolling Patri Friedman's Seasteading Institute, whose aim is to build floating cities similar to oil rigs that would be anchored in international waters and so be free of the laws, regulations, and by extension, the norms and customs of landlubbers (Patri is the grandson of the neoliberal guru Milton Friedman). Perhaps following in the footsteps of Henry Ford's model city, Fordlandia, built in the Brazilian state of Pará, Microsoft purchased twenty-five thousand acres in Arizona to build the City of the Future, while Page has invested in research on the feasibility of privately owned city-states.[26] More recently Jeffrey Berns, a cryptocurrency millionaire, bought a chunk of land in Nevada bigger than Reno to build a Utopian blockchain community. Berns believes that blockchains—decentralized, distributed, public digital ledgers facilitated by autonomous, peer-to-peer networks—will give people rather than companies or governments power: "I don't know why," says Berns. I just—something inside me tells me this is the answer, that if we can get enough people to trust the blockchain, we can begin to change all the systems we operate by."[27]

Blockchain Utopias and the Singularity are pretty far afield from phone apps and social networks, but the expansive simplicity of its vision is what makes the emergent Silicon Valley spirit of capitalism so appealing. The more we share, the more everyone benefits. Turning everything into an always-connected, data-generating device, collecting and storing every bit and byte of data, quantifying every possible piece of life: these practices will not only generate inclusive growth, they'll lead us to the digital promised land. If we trust in these companies, give them free rein to pursue their dreams, they will provide us with new mechanisms to achieve the American Dream and, who knows, maybe even take us to a wondrous future where data and know-how have solved the world's problems.

Uber Unimpressed

And yet . . . The Silicon Valley spirit is only just crystallizing and its appeal is diminishing. It wouldn't be the first time that Americans became disenchanted with technology. Throughout American history technological advance has been coupled with public anxiety and anger over how machines and the companies who wield them are reengineering society. In his classic work, *The Incorporation of America*, the historian Alan Trachtenberg describes the public mood in the late nineteenth century:

> Each act of national celebration seemed to evoke its opposite. The 1877 railroad strike, the first instance of machine smashing and class violence on a national scale, followed the 1876 Centennial Exposition, and the even fiercer Pullman strike of 1894 came fast on the heels of the World's Columbian Exposition of 1893.[28]

A similar characterization could be made of the present. Popular weariness and distrust of Silicon Valley and the technology it is developing are eloquently expressed in works of popular culture: television shows such as *Silicon Valley*, *Westworld*, and *Black Mirror* and novels such as *Whiskey Tango Foxtrot*, *The Circle*, and the uncannily prescient *Super Sad True Love Story*. These pop explorations of how technology is shaping society range from dyspeptic satire to terrified (and terrifying) dystopian depictions of the future, should we continue down our current path. They function as real-time critique. Despite all the mystique surrounding deep learning and the dark web, our pop culture dystopias reveal a society well on its way to articulating a clear critique of what we don't like about the Silicon Valley vision of the future.[29]

Our clarity is in part linked to the peculiar fact that many Silicon Valley visions about what technology should look like, and what our aspirations regarding technology should be, were originally located in science fiction. *Star Trek* fans say the smartphone is the real-life incarnation of the tricorder. Elon Musk, the tech entrepreneur who founded Tesla, named his two spaceport drone ships *Just Read The Instructions* and *Of Course I Still Love You* in tribute to the "Culture" novels by the science fiction great Iain M. Banks. Many classic speculative fiction stories—for instance, *The Winter Market*, by William Gibson, and *Ubik*, by Phillip K. Dick—also provide vivid depictions of the dystopian future technology would or at least could bring. So it's not surprising that we are so adept at creating and indulging in digital nightmare scenarios.

But this fiction-reality overlap is only a tiny part of the mistrust and weariness of the Silicon Valley worldview; it is rooted in something more concrete than our fruitful imaginations. Our disenchantment is rooted in the very real misdeeds of Silicon Valley corporations. It's rooted in the growing realization that, like the titans of the Gilded Age, today's titans mistake their own interests for those of society. Consider Uber, a company that perfectly embodies the Silicon Valley spirit. Uber rose from a tiny San Francisco startup, hiring its first employee in 2010, to a juggernaut in record time. By 2015 it was the toast of the town, seemingly on track to dominate the world of ride sharing. Today it operates in 785 metropolitan areas, has one hundred million users worldwide, and has raised over $22 billion from investors.

Uber reinvented the American myth of success, and not just for its founders. Ordinary folks with a decent vehicle were promised a path to financial independence and the chance to be their own boss. Uber's prestige and popularity made it a desirable place to work for young software engineers and computer programmers. The company promoted an image of a work-hard, play-hard environment where employees were changing the world and having fun doing it. Travis Kalanick, the face of the company when he was its CEO, was a media darling with reporters apt to marvel at his brightly colored sneakers and boyish charm.

Uber's glow faded fast, however, after Susan Fowler, a young software engineer at Uber, clued the world in on what really went on at the company. Fowler—who grew up in an impoverished family, dropped out of high school to help made ends meet, yet managed to teach herself math and earn a doctorate from Carnegie Mellon University—was excited to get a job as a site reliability engineer at Uber. Fowler's story seemed like meritocracy in action, a fulfillment of Silicon Valley's American Dream reboot. But on the first day of work with her new team Fowler's manager propositioned her for sex. Things didn't improve. After a year of failed attempts to improve her work environment from the inside, Fowler quit in 2017 and left a fitting parting gift—a frustrated and powerful blog post detailing the chaos and abuse that raged inside Uber.[30]

Fowler's letter struck a chord. In the months following, story after story poured out about Uber's bad behavior. The company saw a mass exodus of its executives, several lawsuits, and a viral #deleteuber campaign. It was forced to pay $7 million to settle hundreds of claims of harassment, gender discrimination, and creating a hostile work environment. Perhaps most damning after Fowler's missive was the so-called "Jacob's letter," a thirty-seven-page document written by a former security team member, Ric Jacobs, in which he detailed how the company, with the knowledge of Kalanick, used spies and bespoke software to surveil celebrities, politicians, labor unions, cops, regulators, and others; how it hired people to impersonate pro-Uber protestors; and how it engaged third-party vendors to steal data from and sabotage competitors. Uber investor and tech entrepreneur Hadi Partovi summed up the debacle: "This is a company where there has been no line you wouldn't cross if it got in the way of success."[31]

Moreover, the dream of entrepreneurship and making a living as an Uber app driver was just that, a dream. The Federal Trade Commission forced Uber to pay a $20 million fine for misleading drivers about how much money they'd actually earn. When vehicle wear and tear, insurance, gas, and hours spent driving are factored in, Uber drivers struggle to make a living wage. In New York City, an Uber driver named Doug Schifter shot himself in early 2018 outside City Hall in Lower Manhattan after posting a letter on Facebook detailing how indebted he had become; he drove over a hundred hours a week, yet his credit cards were maxed out and he couldn't pay his bills. Schifter's death was one of several suicides by New York taxi drivers in 2018. The unrestricted entry of Uber created a super-saturated taxi market in New York City in which thousands of new drivers were signing up every month, preventing anyone from making a living wage. In the summer of 2018 the city agreed to a twelve-month halt on new driver registrations to give it time to "study the industry," but the damage had been done. When a driver happened to pick up Kalanick for a ride and took the opportunity to raise some concerns about the difficulty of people like him trying to make a go of it, the Uber CEO flew into a rage, screaming at the driver, "Some people don't like to take responsibility for their own shit!"[32]

The mood surrounding the Silicon Valley miracle has shifted. Illustrative of this shifting mood is a recent crime that occurred in San Francisco: unknown malcontents nabbed a surveillance robot outside a pet shelter; the owners had purchased it to keep homeless people away from their storefront. The botnappers threw a tarp over the unsuspecting robot and then disabled it by smearing its sensors with barbecue sauce. Instead of garnering sympathy, the pet shelter's owners were subject to public scorn while the kidnappers were painted as heroes.

Granted, our feelings toward robots are complicated. UC Berkeley students held a candlelight vigil for a food delivery robot that self-combusted. But increasingly the Silicon Valley vision of a future dominated by artificial intelligence and the datafication of everything generates more fear than optimism. Uber's dark side was just one example; every day new misdeeds are revealed. Chris Gilliard is an English professor at Macomb Community College who uses his Twitter account (@hypervisible) as a teaching tool about abuses in the tech industry. Gilliard posted a question to Twitter asking users to cite transgressions committed by tech companies that they knew of. Within a short time users posted over five hundred replies, ranging from familiar grievances such as Samsung's spying television to disturbing reports that women's period tracker data was being sold to aggregators to rumors that Facebook and other companies listen to phone conversations even when people weren't using their apps.[33]

The ready airing of misgivings and complaints in response to Gilliard's tweet signals a growing popular feeling that the Silicon Valley spirit, instead of offering a vision of society that ameliorates the problems of neoliberalism (inequality, precarity, alienation, anxiety) and resuscitates a purer form of capitalism defined by innovation and pluck, seems to be legitimating a cluster of companies who are actually putting these problems in hyperdrive and adding new ones to boot. Instead of reveling in the conveniences and connections brought by our smartphones, society is plagued with doubts about the dystopian future we're creating.

How could we not worry? Every day, ordinary people are presented with fresh evidence of their near-term irrelevance. Stories of self-teaching algorithms, autonomous cars, expert reports predicting the disappearance of at least half the world's jobs in the next couple of decades due to automation and robots abound. It's a wonder we don't stay in bed, scrolling through our feeds, awaiting the Singularity in dignified repose.

Indeed, a chorus of warnings from tech naysayers and handwringers suggests we should be terrified of what a Silicon Valley future holds. Data scientists paint a picture of a future in which algorithms determine everything, and in some accounts this future is now. Companies and state and federal agencies have used algorithms to determine whether someone will get parole, get a job, get a loan, get a raise, get public assistance, be accepted to a school, get fired, or get a promotion. Mathematician Cathy O'Neil says many of these algorithms are "weapons of math destruction" because they are "opaque, unquestioned, and unaccountable, and they operate at scale to sort, target, or 'optimize' millions of people," often with demonstrably negative effects.[34]

Algorithm use in decision making is increasing because algorithms are cheap compared to humans, and they are perceived as unbiased processors of data. These characteristics make them an attractive option for agencies tasked with tough decisions, such as cutting social services for needy community members. A recent report by AI Now, an algorithm watchdog group, gives an illuminating example of the problem: When Tammy Dobbs, a cerebral palsy sufferer, applied for acceptance to an Arkansas program for people with disabilities, the state sent a qualified nurse to Tammy's home, who determined that she needed fifty-six hours of home care a week. In 2016 the state used a proprietary algorithm instead of a nurse to assess Tammy's needs, and her home-care allotment was cut to thirty-two hours a week, dramatically reducing her quality of life overnight. Tammy was one of hundreds of Arkansas residents with disabilities who saw significant reductions in their home-care allotments, with no mechanism to explain or challenge the change.[35]

Our smartphones are a training ground for algorithms. Sometimes the results are great: YouTube knows what music video we'd like to watch next. But increasingly, the results are not great. To name just one example, the photos we post and share are being fed into algorithms to perfect facial recognition software used to reboot the pseudoscience of physiognomy and perfect population control. Instead of a future in which we "solve intelligence and use that to solve everything else," a much darker horizon looms in which the massive amounts of data we generate with our phones is used to feed algorithms that discipline and punish us.

Yuval Noah Harari, a well-known Israeli historian and futurist, makes equally uncheerful predictions. In a scenario similar to that of Kurt Vonnegut's first novel, *Player Piano*, Harari forecasts a world in which more than half the population will join the "useless class" by 2050 as a result of job-crushing algorithms and artificial intelligence: "We are now creating tame humans who produce enormous amounts of data and function as efficient chips in a huge data-processing mechanism."[36]

Funnily enough, Silicon Valley loves Harari. Bill Gates calls his work "fascinating," and Nellie Bowles, a prominent tech writer for the *New York Times*, confirmed the Bay Area's admiration for Harari after she followed him around for a few days on his latest book tour in and around San Francisco. The Silicon Valley elite threw him dinner parties and invited him to their campuses and secret working groups. The love fest is puzzling to Harari, who confessed to Bowles in the lengthy interview that he is uneasy with the praise and wonders if his popularity is because his warnings are viewed as unthreatening. This worries and puzzles him further.[37]

But it's not all that puzzling. Harari's prognostications, rooted in Spencerian notions of evolution, align seamlessly with the techno-determinist world of Silicon Valley elites. They see an independent, wild, driving technology steamrolling society in its path, and Harari tells them it is so. Harari is critical of the political implications for liberal democracy and the power of tech companies, but that's not a problem for the disruptors, who can shrug the critique off and proclaim that their particular company or start-up is "making the world a better place," as Mike Judge, creator of the sitcom *Silicon Valley*, satirizes. Or, if they can't quite manage to shrug it off, they have the resources to join the burgeoning crowd of elite "preppers" buying up land in New Zealand and elsewhere to build escape pads for when the human "chips" go feral.

Accounts like Harari's—and he's only the most popular of a broad field of dystopian futurists—legitimize the deeper story, the harder sell, the story that there's nothing we can do to stave off this coming future. These things we've created—algorithms, platforms, pocket computers, big data—have slipped from our grasp, are beyond our control. In this story we can only try to keep up and help the losers as much as possible.

The Silicon Valley spirit is rooted in this sense of inevitability. It's reminiscent of a familiar trope in neoliberalism that markets are part of nature, beyond the control of humans and their elected officials, and are best left to operate in their free and wild state. The Silicon Valley spirit talks about the future and our participation in it, but rarely our intervention. Intervention (except by a small cadre of "disruptors"), or serious reflection and conversation about the underlying contradictions of our smartphone society, are seen as undesirable or impossible.

Not an Implacable Force

Perhaps in a bid to calm growing hysteria, well-respected scientists—people actively developing algorithms, robots, and artificial intelligence—have become more emphatic of late in tamping down the hype of Silicon Valley futurism, particularly concerning the future of work. Responding to an article in the business press predicting the takeover of work by robots, Rodney Brooks, the emeritus Panasonic Professor of Robotics at MIT, called the claims "ludicrous."[38] Greg Ip, the *Wall Street Journal*'s chief economics commentator, characterized reports of the wholesale destruction of jobs by automation and algorithms "baffling and misguided." Robert D. Atkinson, president of the Information Technology and Innovation Foundation, says, "No matter how many times a purported expert claims we are facing an epochal technology revolution that will destroy tens of millions of jobs and leave large swathes of human workers permanently unemployed, it still isn't true."[39]

Data scientists are also speaking up to assert that algorithms don't take people out of the equation and they aren't unbiased or neutral. Algorithms are designed by people (who have their own biases) and are trained on datasets that themselves often reflect bias and discrimination in their collection and design. For example, until recently if you did a Google image search for "men" or "women" nearly every photo retrieved was of an attractive white person. The algorithm reflected what pictures had been uploaded to the web, the available content, which in turn reflected who is represented in society and who is underrepresented. Google has become aware of the bias being reproduced in its Images algorithm and is taking baby steps toward addressing the problem. Algorithms are not beyond our grasp. They are instruction sets written by people, run on machines created by people, and they can be changed and controlled by people.

The common sense of the Silicon Valley spirit, whether regarding jobs, or algorithms, or whatever, is laced with technological determinism, which, David Noble, a historian of technology, has long argued "absolves people of responsibility to change it and weds them instead to the technological projections of those in command."[40] Technology is designed, developed, and chosen by humans. It's not an implacable force controlled by our alien overlords. People, not machines, are deciding the direction of society, and if we don't like this direction we have the power to take a different route.

Ordinary users and consumers are also guilty of technological determinism in framing the problems of our society and our critiques of Silicon Valley, often because this is simpler. Noble warns against simplistic explanations because they "diminish life, fostering compulsion and fatalism, on the one hand, and an extravagant futuristic, faith in false promises on the other."[41] We can and should tell our own stories about the kind of society we want to live in, technology and all. In doing so we can foster new ways of thinking about our digital-analog future.

But what stories should we tell? How do we deal with the myriad issues raised by our smartphones: privacy, autonomy, addiction, surveillance, precarity, narcissism, commodification, democracy? A growing number of voices are weighing in on these questions, which is a good thing because, unlike the techno-determinists who until recently have received a disproportionate amount of airtime, it assumes that we can do something. If our future is not to be Silicon Valley's dys-/Utopia, then what is it to be? What is our version of Utopia? Or, a more manageable question, what relationship should we have with the technology and corporations that have, largely through our smartphones, come to play a central role in the US and global economy in the past decade? How should we ameliorate the tensions and troubles we've discussed in these chapters?

Thinking about the ills of our smartphone society, articulated in many of the criticisms we've raised, one may feel an overwhelming urge to retreat from our phone worlds. For some our hand machines and the connections they bring seem so deeply corrupting that the only way to restore richness and meaning to our lives is to get rid of our pocket computers, or not to use them in the first place. Nilanjana Roy, author of *The Wildings*, says that her Utopia is the space outside cyberspace: "Almost everything good in my life has come from the disconnected space: the quiet piece of writing, the slow, intimate relationship you build with a landscape, the garden you tend, the long afternoons spent with old and new friends."[42] This quest for quiet resonates. Harari spends two months a year in silence at an Indian ashram, while Andrew Sullivan helped kick his internet addiction by checking in to a silent retreat.

Some practice partial disconnection by trading in their smartphones for "dumb phones," or limiting their use through dashboard apps or plain old personal resolve. Families have adopted the practice of "technology shab-bats," disconnecting for part or all of their weekends. Disconnection is par-ticularly popular in the belly of the beast, Silicon Valley. Elites send their children to tech-free schools, such as the Waldorf School, where screen time is taboo, and force their nannies to sign "no-phone contracts," which promise "zero unauthorized screen exposure." Nannies who break the rules, perhaps sneaking in a text to their own children while out and about with their charges, might end up fired, or in a "nanny spotted" post on a local parenting messaging board.[43]

The impulse to retreat from technology—the "exit" in influential econ-omist Albert Hirschman's classic "exit or voice" schema—has echoes that go back to the American transcendentalists. Ralph Waldo Emerson, Mar-garet Fuller, and Henry David Thoreau railed against conformity, advocat-ing independence and the development of intuition. A central part of their worldview was a deep skepticism toward technology. In *Walden*, Thoreau dismissed the significance of the train and the telegraph, calling them "im-proved means to an unimproved end." "We are in a great haste to construct a magnetic telegraph from Maine to Texas," Thoreau declared, "but Maine and Texas, it may be, have nothing important to communicate." In the 1970s, a period of deep crisis, people joined communes or became farmers in the "back to the earth" movement. This vision of retreat is thriving today. Wist-ful articles detail the staying power of the Amish way of life, while YouTube is full of videos of people talking about their van life, or tiny home, or off-the-grid lifestyle. (Interestingly, most people who pursue a minimalist way of life, at least the ones we see on YouTube, seem to get rid of nearly every-thing *except* their smartphones.)

As a personal or a family strategy, to retreat and get rid of or strictly limit phones as a way to deal with the issues raised here is perfectly accept-able. It's a decision that fits neatly within a broader adoption of "lifestyle politics"—aligning personal consumption choices with politics—as a strat-egy for resisting ecological destruction, inequality, personal alienation. There is something very appealing about an imagined phone-free life. We'll be more focused and protected in our personal lives and we'll help the planet by reducing the cost of extraction to make and use the phone and to store all the data we generate.

But a personal strategy is not a spirit. If we want to create an alternative vision for the kind of society we want to build to counter the dominance of the Silicon Valley spirit, we should be cautious about advocating technological retreat. As a broad-scale approach, retreat perpetuates, or at least leaves unexamined, questionable ideas about the relationship between technology and society, such as the idea that humans "before" technology were more pure, and more important, the idea that social relationships are shaped by technology much more than they are by power inequities.

Technology plays a part in power imbalances, to be sure, and technology itself can engender certain political and social outcomes. Langdon Winner argued, for example, that nuclear power plants necessitate authoritarian managerial models because they rely on the deployment of very hazardous materials.[44] But social relationships, such as class, race, and gender divides, are not reducible to technology. Smartphone design doesn't demand that Congolese miners be paid a starvation wage, or that Facebook make psychologically toxic software designed to steal our data. These are political relationships that arise from the way society is organized in our for-profit system. Getting rid of our phones does nothing to change this.

Moreover, advocating retreat obscures the social relationships and power dynamics between users. Those who suggest retreat often have a broad range of choice and freedom when it comes to technological access, but sixty-five million Americans rely on their smartphones as their only access to the internet.

Minu Thomas and Sun Sun Lim's research on Indian and Filipino women working as live-in maids in Singapore highlights a different element of these power dynamics. The maids' phones "foster emotional links with friends and family, grow their social networks and afford them greater autonomy in seeking better job opportunities," which has pushed Singaporean employers to limit or ban phone use by their maids.[45] When Valley elites demand that their nannies eschew phone use during the workday, a similar power dynamic is at play. They cut off these nannies' access to their own family during the long hours they work, reduce their ability to talk about their working conditions with other nannies, and increase their isolation. A strategy of retreat reinforces these divides.

By extension, retreat as a societal strategy relinquishes control. It encourages people who are critical of our emergent smartphone society to step out of the conversation about what kind of world we want with technology in it—deciding what the priorities should be. It leaves these questions to be decided by whoever is left, which in practice means the private corporations who are already in control, and who, as history has shown, have little incentive or propensity to regulate themselves. The fact is smartphones aren't going anywhere. Like the automobile they've become integrated into life. We have the power—if we take it—to shape what that integration looks like.

Jaron Lanier, a virtual reality pioneer, is on the retreat bandwagon, but he advocates a more circumscribed retreat. He urges us to teach companies like Facebook and Google a lesson by becoming social media refuseniks until they agree to give us back control over our data. Only when they agree to pay us for our data, Lanier argues, should we agree to return: "I won't have an account on Facebook, Google, or Twitter until . . . I unambiguously own and set the price for using my data, and it's easy and normal to earn money if my data is valuable. I might have to wait a while, but it'll be worth it."[46]

Eric A. Posner, a Chicago law professor, and E. Glen Weyl, a senior researcher at Microsoft Research, also advocate ownership. They argue that we are all digital laborers now and we should be getting paid accordingly. Granted, we might not get paid a lot initially, Posner and Weyl argue, but "that amount could grow substantially in the coming years. If the economic reach of AI systems continues to expand—into drafting legal contracts, diagnosing diseases, performing surgery, making investments, driving trucks, managing businesses—they will need vast amounts of data to function." As a bonus, Posner and Weyl contend, selling our data could make up for job loss resulting from artificial intelligence—we'll earn a wage from all the things we do for free now on our apps and platforms.[47]

The idea of becoming owners of our own data with the right to sell it as we see fit has widespread appeal. Like the dream of retreating from technology, the idea has deep roots in American ideals. Bruce Schneier articulates this lineage in his critique of the business model used by companies such as Facebook and Google: "We are tenant farmers for these companies, working on their land by producing data that they in turn sell for profit."[48] The tenant farmer–landowner dynamic is reviled in American mythology, seen as an affront to the ideal of the yeoman farmer who owned his own land and was honest, hardworking, and independent. The vision of owning and selling our own data transplants the warm feelings we have about yeoman farmers to the present. It's an empowering vision in which each of us owns and sells what we produce. It is also a vision that restores the imagined promise of early American capitalism, a place where virtuous, propertied citizens trucked, bartered, and sold in a competitive marketplace defined by choice and equal competition.

The practicability of an ownership approach in which someday we might support ourselves on the data we produce is questionable, given how cheap our data has become. Platform companies make money through economies of scale, because they own and control access to so much data, not because the data each of us produces is particularly valuable. Right now Facebook and Google only make between ten and twenty bucks a year per user, and data is getting cheaper by the minute. So the real appeal of a yeowoman smartphone society is not in the price we could potentially command for our data but rather the *choice* to sell or, more important, not sell one's data at all.

Thinking about the question of choice, however, highlights a deeper limitation of the ownership vision. By agreeing that the moments of our lives, interactions, inner thoughts, pictures, videos, and daydreams should be quantified and commodified, we give up something. We turn ourselves into microentrepreneurs of life. People have long fought, and continue to fight, to keep spheres of life—our families, our religions, our schools, our community institutions—outside the market, and we shouldn't stop now. Most of us don't use our phones for mercenary purposes but to find joy, information, and connection. If we were to apply a market logic to these ends, to start thinking about how to make the most money, the most efficiently, from our smartphone lives, where does that end? Our interactions are not commodities; life is not a production process.

Right now our phones are being used to obscure and perpetuate an unequal relationship, designed to favor a few large companies. It's not a difficult situation to understand once we blow away the metaphysical mist surrounding our feelings about technology, human nature, God, existence, and so forth. These two visions for a different kind of society—retreat and ownership—respond to popular critiques of Silicon Valley and fit within the broader canon of American ideals. But they are also limited visions.

In this political moment there is a broad struggle over meaning, and as a result, an opening for a different vision. Instead of the Silicon Valley spirit, or visions focused on delinking or commodification, we can construct our own spirit, our own set of ideas about how we want our smartphones, and technology more broadly, to be incorporated into society. Now is the moment to make big demands that establish the horizon of the society we want to build. We don't know what this vision will look like yet—it isn't something that will come from a book. But here are a few principles to consider.

Principle 1: Our Phones Shouldn't Be Used to Perpetuate
and Obscure Coercive and Unjust Relationships.

Many of the critiques raised in these chapters stem from the myriad ways our smartphones are used to perpetuate and obscure coercive and unjust relationships. Our pocket computers are used in ways that reconfigure, and often reinforce and renew, existing power inequalities. We see this power inequality most starkly in the relationship between corporations and workers, and between corporations and consumers. In the gig economy our hand machines mediate the employment relationship, encouraging app workers to view their phones as their boss. But our phones are not the boss; companies are. Right now many high-tech companies are flagrantly violating existing independent-contractor laws. We need to radically update our laws regarding employment relationships for the smartphone age to give workers the dignity, pay, and protection they deserve. The relationship between tech companies and consumers needs an even bigger overhaul. Companies shouldn't have the right to either collect and/or sell our data or access to our data, and the data broker industry is ripe for elimination. At the very least we should have the right to easily delete *forever* any text, videos, or photos that appear about us on private platforms.

These demands are part of a much broader discussion about our right to privacy. Privacy is a fundamental right that needs to be reemphasized and reconsidered in the smartphone age. Instead of privacy being an afterthought, or worse, purposefully violated, in the design of our technology its importance should be written into the design of the software, algorithms, and apps that make the modern world go round. Free-software pioneer Richard Stallman argues that we need to redesign computer systems to have limited data collection ability, not just to regulate how the data is used once it is produced: "If we really want to secure our privacy, we've got to stop the collection of the data. Rules to limit how the collected data may be used may do some good, but they're not very strong protection. The privacy issue is broader."[49]

The question of privacy also highlights the role of governments, especially the US government, as major purveyors of coercive and unjust behavior. Governments shouldn't have the right to monitor our every keystroke, swipe, and tap, to take photos and video of us to use with facial recognition software. They shouldn't be able to surreptitiously use this data to monitor, harass, arrest, and even kill, either in the United States or abroad. This includes local police departments that use social media to quash civil disobedience and community organizing. It feels daunting to demand a stop to a pattern of government surveillance that seems to know no bounds, but it can be done. The US government was forced to rein in its surveillance practices after the 1975 Church Committee revelations of illegal CIA surveillance of groups and individuals. With concerted effort it can be forced to again.

Finally, our phones are also used to perpetuate and obscure unequal power relationships between people. If we're serious about restoring privacy, we need to get serious about enforcing repercussions for those who publicize information shared in confidence in an attempt to humiliate, discredit, or otherwise harm another person. For example, instead of shaming teenage girls who sext, we should hold the people who publicize private messages accountable. At the same time, people shouldn't be able to hide behind their apps and platforms to bully and harass others. We need to have a public conversation about free speech and what we expect from the applications that we use. Twitter, Facebook, Google, and other websites are not democratic organizations run in the interest of the communities they serve; therefore it shouldn't be up to them to decide or mediate the shape and content of public speech.

Principle 2: We Shouldn't Use Our Phones to Mask Bad, Selfish, or Immoral Behavior.

One of the most amazing things about our phones is how they empower us as consumers. With a tap and a swipe we can beckon life-easing services and access a mind-boggling range of consumer goods. Indeed, it's so easy to consume that our phones often facilitate unthinking consumption. But Americans should think hard about our consumption norms because they are underwritten by privilege—for example, the power that comes from elevated social status or living in a wealthy country—and this privilege often comes at the expense of others.

We may be able to get a head of broccoli delivered from Postmates in an hour, but this doesn't mean we should. We can make do with what's in the pantry or walk to the store. The new consumption norms encouraged by our smartphone economy are not ecologically sustainable, and although we may not see the ancillary cost when we tap our apps, there is a price to be paid. This doesn't mean we should depend on lifestyle politics as a solution to climate change and habitat destruction. It means that as a society we must have concerted conversations about the kind of future we want. What do we value more, millions of products available on demand and a private car that arrives in three minutes, or a planet that's suitable for human life?

Just as our phones hide unequal relationships, they also can encourage bad behavior. We witness and participate in this behavior when we watch a video of someone overdosing on fentanyl, a video shot by a bystander who's filming rather than helping the person or the screaming toddler standing next to them. The internet abounds with evidence of bad phone behavior—outing nannies who dare to check their phone while working, real-time photos and comments about unsuspecting plane passengers taken surreptitiously by the person sitting behind them, virtue signaling pile-ons, and infinite varieties of trolling.

Our bad phone behavior isn't usually a matter for the police. It is evidence of a society whose norms are in flux. It's time we develop new social norms for the smartphone age to protect people's dignity and privacy, as well as our own. Developing new social norms about acceptable smartphone behavior will enable us to update our concept of privacy, a crucial objective that will give us standards for acceptable behavior and practices, whether between ourselves and our own device, between people, or between institutions and people. It will also encourage more reflexivity about our own motivations and agendas when we use our phones, a necessary process if we're going to get control of the machine in our pocket.

Principle 3: Our Phones Should Be a Pathway to a True Digital Commons Where Life Isn't for Sale.

After two shouldn'ts here is a should. The internet is an interesting, hilarious, and occasionally wondrous place. It is a great human invention. So are smartphones. The 1999 report "Funding A Revolution: Government Spending for Computing Research" demonstrates how the federal government played a central role in building the nation's computing and communications infrastructure through taxpayer funded research and development.[50] Mariana Mazzucato points out in her award-winning book *The Entrepreneurial State*, "Apple was able to ride the wave of massive State investments in the 'revolutionary' technologies that underpinned the iPhone and iPad: the internet, GPS, touch-screen displays, and communication technologies."[51] Moreover, our unpaid, appropriated work is what makes these platforms so valuable and central to modern life. As such, we have the right to demand what kind of digital society we want—the right to demand that our smartphones be a pathway to a true digital commons, a place where our data, inseparable from our life, is not for sale at any price.[52]

One of the greatest tensions of our smartphone society is the disconnect between the motivations shaping how ordinary people use their phones and the motivations of the tech titans who control our phones. One group is interested in socializing, learning, doing politics, and being entertained, while the other is primarily interested in making a profit through extracting user data to sell to other companies. These conflicting drives are ultimately irreconcilable. Omaha farmers captured this underlying contradiction in the first Gilded Age: "We believe the time has come when the railroad corporations will either own the people or the people must own the railroads."[53] We're at a similar crossroads. If we want expanded privacy and dignity as technology advances today we need to create spaces that allow for privacy and dignity. Right now the platforms and apps that both individuals and society have come to rely upon are not those spaces.

Does that mean we should build a People's Facebook or Google? Perhaps. It might mean running these companies as utilities, recognizing their centrality to everyday life and thus regulating them as such. Or it could mean advocating that every person receives a digital baby box, riffing on Finland's practice of sending expectant mothers a box with everything she needs for her new baby. Instead of diapers and onesies, each person would be given an encrypted email account, access to taxpayer funded broadband internet, secure server space, and a library of web and social media tools that follow free-software principles and don't collect and sell people's data.[54]

In the present climate goals like these might seem out of reach. But the barriers to achieving them are not technological or financial. The digital architecture that undergirds our smartphone society is surprisingly lean in terms of manpower and capital costs. Recall that Instagram only employed thirteen people when it was acquired by Facebook. Moreover, the know-how to build a public system subject to democratic design and control is widely available. The biggest barrier today is the "common sense" cultivated by Silicon Valley that our country's digital architecture should be discussed and decided by the people who own the tech companies rather than by the ordinary people who use the technology, provide the data and the taxpayer funding, and are impacted so profoundly by the priorities and proclivities of the tech titans.

Instead of giving in to techno-determinist impulses designed to keep us in our place we should encourage a broad and hearty debate about what a true digital commons should look like. It should be a debate that's not afraid to make big demands and isn't dominated by software engineers and elites. Our smartphones have brought digital technology into the most intimate spheres of life. It's time to take control of them and repurpose them as pathways to a democratically designed and maintained digital commons that prioritizes people over profit.

Chapter 6

New Map

Smartphones are peculiar creations—their guts locked up tight, they hide deep secrets. Their diminutive stature and smooth expression tend to obscure how the technology, institutions, and relationships they embody are a driving force in our twenty-first-century society. But if we pause and take a closer look it is clear that our smartphones also have a special power to reveal these mechanisms. Our pocket computers are a microcosm of modern life. Examining the technology, institutions, and relationships embodied in our smartphones, as we have done in this book, facilitates what Frederic Jameson calls cognitive mapping—a process by which each of us situates ourselves in the social, political, and economic totality of the contemporary moment.[1]

In our emergent smartphone society people are connected by layers of intertwined networks, both digital and analog. These networks are conduits for rapid information flow, measured in the trillions of texts, videos, posts, and likes that spread ideas and new cultural norms. Participating in these networks lets individuals feel part of something bigger than themselves—a community, a social movement, a zeitgeist. Many connections are weak and exist primarily in the digital realm: Facebook "friends" or Instagram followers that one "sees" but never meets in person—"familiar strangers," as Stanley Milgram might call them.[2] However, overlaid onto this thicket of weak ties are much stronger paths of connection between analog friends and family. Almost universally, people use their pocket computers to solidify and maintain bonds with the people closest to them.

Our cognitive maps of our smartphone society are also maps of capitalism. Corporations engage in territorial conquests of cyberspace, and cement their reigns through privileged access to capital, a lax regulatory structure, and an ideological environment that delegitimizes democratic control. The nature of corporate control is always evolving; today companies are using smartphones, along with digital technology more broadly, to develop new profit-making models. As part of a longer shift in the organization of work, tech companies are using apps and spin to rework work, reengineering it as a space in which workers are paid for tasks rather than time, and workers, consumers, and governments shoulder the brunt of production costs rather than companies.

The tech titans are also quietly, stealthily, making grabs for territory in previously uncommodified spaces. Our phones have blurred and reconfigured the ever-shifting boundary between the public and private. A key part of this reconfiguration is how our new tether creates a constant stream of data—geographic, behavioral, psychological—that corporations and governments gorge on. They appropriate our work while framing our unpaid content creation and data generation as a labor of love. All the data generated through these new digital networks and the information people ingest and expel through their umbilically attached hand machines mark a new frontier, a place capital is just now exploring and exploiting. Eager to capture windfall gains from commodifying new spheres of life, the titans conquer as they go, creating software and algorithms to transform conversations, meanderings, performances, and desires into dollars and market share.

In this campaign companies have a partner in governments who engage in ubiquitous mass surveillance in order to monitor and suppress pockets of digital-analog dissent. US intelligence entities bend geolocation software to geopolitical ends, using our phones to transform the planet into a seamless battlefield. Battlefield technologies make their way home through local law enforcement and government agencies that use social media, bespoke software, backdoors, and massive server farms to monitor, record, and store our taps and swipes.

Turbulence is unavoidable in any system, not least in capitalism; the terrain of our smartphone society is not a placid landscape. It is a contested terrain, particularly in the past decade. Since the 2008 financial crisis individuals and grassroots organizations on both the left and the right of the political spectrum are challenging the putative "common sense" of neoliberalism. Our hand machines have become weapons in this contestation. Demands for justice and security by ordinary Americans are part of a constellation of progressive digital-analog political movements that have emerged in the past decade. In these movements folks are using their social networks and constant access to information to demand a redrawing of the map, to take back control of digital and analog territory ceded to corporations and the state.

Our cognitive maps reveal the fascinating dynamics that have emerged in the past decade. But they are incomplete, reminiscent of the maps drawn by ancient cartographers with fantastic beasts, wild disproportions, and blank spots where geographical formations should be. Today's misrepresentations and blank spots are our blindness to key elements of the digital frontier. The contours of the big-data landscape are blurry and incomplete. We can't see the algorithms, software, hardware, supply chains, energy, infrastructure, and people—the "sociotechnical layers"—that underlie the making of this new frontier. The mineworkers, assemblers and dissemblers, and content moderators along with so many others who make our phone worlds possible exist outside of the average user's frame.

Other lacunae are the ecological externalities smartphones produce. Our inability to comprehend the ecological dimensions of the world we create is not new, but our hand machines seem to increase the size of these blank spaces rather than diminish them. Happy stories and glossy screens direct our attention away from a frightening reality of ramped-up consumption, perpetuating a widespread obliviousness to how new technologies governed by the rules and norms of our for-profit economy are compounding global ecological devastation.

Jameson's concept of cognitive mapping is not just about learning or knowing as an end in itself, however. Cognitive mapping is about understanding power in order to locate entry points of struggle. By pinpointing and revealing the contradictions of our smartphone society, our hand machines themselves become sites of struggle. Today we're witnessing growing resistance. The deep crisis of legitimacy plaguing neoliberal capitalism has sparked new eruptions of long unresolved struggles over racism, sexism, inequality, and environmental justice. But also growing are struggles more closely linked to the issues raised by our smartphones: precarity, privacy, and the commodification of everyday life. A wide range of groups is working on these issues right now, actively challenging norms and redefining our relationship to smartphone and digital technology. In fighting these fights they are charting a path toward a more equitable society.

Pushing Back

In the future envisioned by Silicon Valley machines make the world go round and people fade into the background as passive consumers. But this future is a fantasy. Tech companies use a wide range of strategies to devalue, deemphasize, and disappear the human labor, most of it low paid, that they rely on. They outsource and hide the people who create training data, moderate content, clean their campuses, and drive their shuttles. They use appwashing and gamification to disguise shit jobs as entrepreneurial opportunities, and they apply peer pressure and nondisclosure agreements to chill dissent among full-time employees. But as the disconnect between fantasy and reality becomes too obvious to ignore workers are fighting back.

In 1998, long before people were using apps to connect with "transportation network companies," New York City taxi drivers formed a union, the New York Taxi Workers Alliance. The taxi market was widely considered impossible to unionize, but under the bold leadership of Bhairavi Desai, and through clever tactics and dogged organizing, drivers won important gains: regulations on taxi companies, a health and disability fund for drivers, and raises that put drivers on the path to a living wage. But the arrival of Uber and Lyft threw a wrench into the works, "disrupting" not only the taxi market but also the lives of taxi drivers. Debt and driving time rose while compensation and security plummeted. Where once taxi drivers could make enough to buy a home and send their kids to college, today 85 percent of app drivers earn below the minimum wage.[3]

But drivers are fighting back.[4] The New York Taxi Workers Alliance was instrumental in getting the city to put a cap on new driver regulations (the first in the country) in August 2018, and a month later the union submitted a multipronged proposal to the New York City Taxi and Limousine Commission calling for a raise for all drivers, including app drivers, a debt relief program for medallion owners, a minimum app fare so drivers make at least $10 a trip, regulations that would align driver pay with changes in Uber's pricing schemes, and retirement benefits for all drivers.[5]

App drivers aren't the only workers demanding recognition and fair compensation for the work they do. Working Washington, a statewide workers organization, recently forced Instacart to admit that it was stealing its delivery workers' tips, using customer tips left through the app to fulfill the $10 minimum payment it guarantees its drivers, rather than paying the minimum itself.[6] Under pressure from delivery drivers, DoorDash has admitted that it follows the same practice; CEO Tony Xu recently announced that the company will stop the practice and promised that in the future drivers' earnings "will increase by the exact amount a customer tips on every order."[7]

On Silicon Valley "campuses," employees can bowl a game, get a haircut, have their clothes washed, and get it all done for free. Yet the low-paid service workers, 60 percent of whom are Black or Latino, who make these wonderlands go are expected to fade into the background. Nothing demonstrates who is valued and who isn't better than the pay disparity between full-time, permanent employees and "blue-collar," contract tech workers— workers who don't receive the pay or benefits enjoyed by full-time employees despite filling critical roles ranging from content moderators and customer support reps to janitors and shuttle bus drivers. In Santa Clara County blue-collar tech workers make an average of $19,900 a year whereas permanent, full-time employees of the same companies make $113,000.[8]

Blue-collar tech workers have had enough. Facebook cafeteria workers voted to unionize in March 2018. Facebook also agreed to pay all its contract employees $15 an hour, following a broader nationwide push by the Fight for $15 campaign. In May 2019 the company announced that it would increase the minimum wage for contract workers in the Bay Area, New York, and Washington, DC, to $20.[9] Shuttle drivers at Loop Transportation and Compass Transportation, whose buses ferry workers at Apple, Yahoo, eBay, and others, voted to unionize, and security guards at Apple and Google won recognition as full-time employees. The tide is turning. Over the past few years, over 5,000 tech service workers have voted to unionize.

Many blue-collar tech workers are organizing through Silicon Valley Rising, a community-, faith-, and labor-based coalition created to fight occupational segregation and income inequality. SVR sees the struggles of tech workers as rooted in the broader dislocations and disruptions caused by tech companies' rapid and unrestrained invasion of Santa Clara County over the past few decades. The coalition calls not only for wage gains but also affordable housing, and an inclusive tech industry; right now only one in ten permanent employees of tech companies are Black or Latino.

Low-paid tech workers are fighting for bread and butter, but they aren't the only workers in the tech industry who are fighting. Recently well-paid workers have also been standing up to the bosses, belying their position as the "labor aristocracy," or at least suggesting that a coalition might be possible. Google employees have been particularly vocal. Many were disturbed and began organizing when they learned about a secret company program called Maven to provide artificial intelligence to the US military to improve the speed and accuracy of sorting drone footage and photographs. In co-ordination with the Tech Workers Coalition, a recently formed group of tech workers and community and labor organizers, employees generated an internal company petition that garnered over four thousand signatures stating: "We can no longer ignore our industry's and our technologies' harmful biases, large-scale breaches of trust, and lack of ethical safeguards. These are life and death stakes."[10]

The company was surprised by and initially dismissive of the petition. When, at an all-hands meeting, a woman raised the point that she had left the Department of Defense so that she wouldn't have to work on these types of projects, and she asked what voice Google employees had to protest decisions like these, Sergey Brin responded that her right to ask the question was the voice she had.[11] Apparently that wasn't the response workers were looking for. After months of organizing, including an open letter signed by dozens of respected scientists and public figures criticizing the ethics of the Maven project, Google backed out of the contract. The victory emboldened tech workers at other companies. At Amazon, employees lobbied executives to stop selling facial recognition software to law enforcement, while tech workers at Microsoft and Salesforce called for their companies to cancel contracts with Immigration and Customs Enforcement.

Google workers have maintained their momentum since the Maven victory. They've walked out over Google's cover-up of sexual harassment, protested the company's plan to build a censored search engine for the Chinese government, and critiqued Google's tiered workforce structure. Another bone of contention was Google's policy of forced arbitration: in workplace disputes employees are far less likely to win in forced arbitration than they are in state courts. Worker actions led to Google's concession to eliminate this policy.[12] Several other tech companies (Facebook, Airbnb, eBay, Square) have since announced they'll also ditch forced arbitration.

While labor coalitions are burrowing from the inside, with a focus on improving the lives of tech workers and their families, other groups are organizing on a larger scale, setting their sights on the ways digital tech—particularly tech popularized in the past decade—is impinging on Americans' civil rights. The American Civil Liberties Union has tackled issues of privacy and technology head-on. At both the national level and in its local chapters the ACLU is focused on myriad internet-related issues, including internet privacy, surveillance technologies, location tracking, and consumer and workplace privacy. It has sued the US government for information on government hacking[13] and FBI social media surveillance, and has filed numerous FOIA (Freedom of Information Act) requests, including a recent demand for information on how the Department of Homeland Security is using Amazon's Rekognition facial recognition software to target immigrants.[14] The watchdog group was also instrumental in the passage of a recent ruling prohibiting state officials from blocking users from official Facebook pages.

While the ACLU fights civil liberty violations with lawyers, ProPublica uses journalists—a pack of sleuths who "expose abuses of power and betrayals of public trust by government, business, and other institutions, using the moral force of investigative journalism to spur reform through sustained spotlighting of wrongdoing." These days ProPublica has been investigating the tech titans, publishing an in-depth story on age discrimination at IBM (which has led to a lawsuit), a how-to guide to finding out what information data brokers have on you, and a series of projects examining algorithmic injustice.[15] The group is responsible for publicizing how Facebook's ad tools enable purchasers to get around federal antidiscrimination laws. It has also been building a database on targeted political advertising. To gather data ProPublica created a plug-in, installed voluntarily by users, that automatically sends the political ads shown to the user along with their targeting information (age, location, political views, gender, and so forth) to the database. Over 22,000 users installed the plug-in and collected more than 120,000 ads. Facebook demanded that ProPublica stop the data collection; when it refused to do so, Facebook rewrote its website code to disable the ProPublica app, demonstrating the power of tech companies to control the digital terrain.[16]

The Institute for Local Self-Reliance shares the ACLU and ProPublica's David and Goliath sensibility in the fight over digital divides, describing its mission as "challenging concentrated economic and political power," and championing "broadly distributed" ownership, "human scaled" institutions, and "decision-making that is accountable to communities." But instead of focusing on blocking digital access—for example, the unconstrained access of companies and governments to people's data—ILSR is fighting to improve access to the internet for poor and rural communities. Imagine being unable to watch a YouTube video, or having to wait a day for an email to be sent, or having no access to the internet at all. This is the reality for the one quarter of rural Americans who have no access to high-speed internet. One of ILSR's major projects to ameliorate this situation is its community broadband initiative, which seeks to dilute the power of internet provider monopolies (cable and DSL), which have the power to determine the provision and price of internet services. So far the grassroots movement has seen more than eight hundred communities invest in some kind of public internet option, with roughly five hundred forming municipal networks, and about three hundred forming cooperatives.[17]

In line with its community focus, ILSR has also taken on Amazon, in particular its marketplace model. ILSR, along with researchers at the Open Markets Institute, argue that Amazon's e-commerce monopoly forces small businesses to use its marketplace platform; through detailed case studies they show how Amazon's structure and business model makes small companies vulnerable to fee gouging and copycat underselling from the "everything store."[18] ILSR is part of a growing chorus calling for rebuilding robust government antitrust laws to protect consumers and communities in the digital age.

Although we often think of the tech titans as American companies, and most of them are, their power extends far beyond the United States. In Europe, Google, Facebook, Microsoft, Amazon, and Apple dominate the tech landscape. However, if a string of recent decisions led by the EU's Commissioner for Competition, Margrethe Vestager, has anything to say about it, it would seem that Europe has had its fill of Silicon Valley rule. Since taking office in 2014 Vestager has ordered Apple and Amazon to pay millions in taxes, has handed Alphabet and Google multibillion-dollar fines for breaching antitrust rules, and has given Facebook a slap on the wrist for misinforming customers about the nature of its takeover of WhatsApp.

More sweeping is the EU's new privacy directive, the General Data Protection Regulation, which went into effect in May 2018. The GDPR regulations require that any company or website collecting personal data must gain user consent to use the data and must say why it is collecting the data, whether it is sharing it, and how long it plans to keep it. Data collectors must also provide users with a portable copy of the data and in some cases grant users' requests to delete personal data. Companies such as Google and Facebook that deal extensively in data mining and selling access to data must hire a data protection officer and report any data breaches that impact user privacy within seventy-two hours. Violators of GDPR regulations face hefty fines.[19] Also in 2018, California passed the California Consumer Privacy Act, allowing individuals to know what data is being collected about them and to delete data and restrict its sale.

The GDPR is one policy approach to a sweeping set of problems—the incursion of tech companies and their software and algorithms into the nooks and crannies of daily life. The GDPR's creation originated in complaints by a former law student, Max Schrems, about Silicon Valley firms' violations of European privacy laws. Other advocates for digital justice point to how tech companies do more than just invade our privacy—they develop and deploy algorithms that can cause substantial harm to individuals and communities. A growing number of data scientists advocate shining a light into the "black-box algorithms" that are being rapidly integrated into decision-making processes in myriad spheres of life. They call for "algorithmic accountability," which the nonprofit research institute Data & Society defines as "the assignment of responsibility for how an algorithm is created and its impact on society; if harm occurs, accountable systems include a mechanism for redress."[20]

Nicholas Diakopoulos, director of the Computational Journalism Lab at Northwestern University, and Sorelle Friedler, a computer science professor at Haverford College, suggest five dimensions of algorithmic accountability: responsibility, explainability, accuracy, auditability, and fairness. Essentially, Diakopoulos and Friedler say there should be a person in charge to deal with problems that arise and be able to fix or override the algorithm; the algorithm's design should be explainable to the people being impacted by it; data errors and statistical uncertainty are unavoidable so we need to systematically measure the accuracy of algorithms; public auditing is a must and would lead to more conscious design; and finally, the social results of algorithms should be evaluated to ensure they are fair and don't reproduce discrimination.[21] Advocates for algorithmic accountability have adopted a strategy that zeroes in on the nuts and bolts of the technology to address societal problems.

Proponents of another policy initiative, the Green New Deal, are tackling the ecological implications of our smartphone society. Proponents of the Green New Deal embed recent global technological shifts, which are highly energy- and resource-intensive, within the context of broader climate trends. As we have discussed in these pages, many of the issues connected to our ubiquitous smartphones have ecological dimensions that are often ignored or downplayed. The Green New Deal encourages us to confront difficult questions about the sustainability of our new digital-analog lifestyles.[22]

Supporters of the Green New Deal recognize that energy consumption is vital to modern society. We need light and heat; we need transportation;

we need computers and industrial machines. However, we also need to decrease carbon dioxide emissions by 40 percent by 2050. Green New Dealers say it's possible to marry these conflicting needs by decoupling energy consumption and economic activity from fossil fuel consumption. This "absolute decoupling" will require a worldwide investment program in which countries collectively direct substantial resources—1.5 to 2 percent of global GDP every year—to boost energy efficiency and produce a lot more clean and renewable energy.

In conjunction with action from the Sunrise Movement, US congresswoman Alexandria Ocasio-Cortez and US senator Ed Markey introduced a version of the Green New Deal in Congress in February 2019. Their resolution calls for the United States to take a "leading role in reducing emissions through economic transformations" because it has "historically been responsible for a disproportionate amount of greenhouse gas emissions." Instead of promoting a "degrowth" philosophy—reducing fossil fuel use by reducing consumption and production—proponents of the Green New Deal advocate a "just transition" toward a green energy ecosystem in which people and communities, particularly those whose livelihoods depend on the fossil fuel industry, are given the support they need during the transition process in the form of job transfers, pensions, and retraining.[23]

Taking Stock

These initiatives, ranging from labor organizing to watchdogs to green policy, are by no means an exhaustive accounting of the current pushback against the ills of our smartphone society. Instead they give a sense of the breadth of the scope and priorities of people and organizations actively working on issues of digital justice. These groups, through their research and advocacy, are charting a course toward a more equitable smartphone society. Examining how the resistance building today meshes with the broader principles we emphasized in the last chapter—that our phones shouldn't be used to perpetuate and obscure coercive and unjust relationships or to mask bad behavior, and should be a pathway to a true digital commons—helps us to take stock of the horizon we're moving toward and also to suggest adjustments to our course.

Many groups' work deals with the first principle: that our phones, and the technology, institutions, and relationships they embody, shouldn't be used to perpetuate and obscure coercive and unjust relationships. Worker coalitions are shining a light on the invisibility and precarious working conditions experienced by tech workers and people working in the app economy; ILSR is showing the importance of access and how rural and poor communities are under the thumb of giant telecom companies; groups advocating for algorithmic accountability emphasize how code can be coercive; and the ACLU and ProPublica are fighting against unjust surveillance and are clear advocates for a broader restoration of privacy.

Seeing the great work these groups are doing also highlights the uphill battle we face. The New York Taxi Workers Alliance is organizing tirelessly to mitigate the hardships of its workers through small but meaningful gains. Yet the notion of compelling transportation network companies to treat app drivers as employees with rights, thus undercutting the model of digital piecework, is fiercely contested. The defeat in court of Deliveroo drivers in the UK who pursued employee rights and the April 2019 National Labor Relations Board advisory memo designating Uber drivers as independent contractors with no right to unionize show how entrenched the new app-work models have become in just a short time, and how difficult it will be to change expectations about work in the gig economy. But difficult does not mean impossible. As this book went to press, California legislators approved a landmark bill that forces app-based service companies (such as Uber and Lyft) to reclassify their workers as employees rather than independent contractors.

Tech companies are sitting atop mountains of cash thanks to mass quantities of unpaid and underpaid work, a technological infrastructure that was developed with taxpayer money, and access to cheap credit for development and expansion, courtesy of low-interest rates engineered by the Federal Reserve. The resources to provide a decent livelihood for all tech workers, whether they're app drivers or content moderators, are available. Opportunities for organizing workers in tech and related industries are also there. When the assembly line was popularized in the early twentieth century observers thought the deskilled work and high-pace environment of factories would never support empowered workers.[24] They couldn't have been more wrong. American manufacturing workers created the labor movement—the most successful social movement in US history, a movement that transformed workers' role on the assembly line from a weakness into a strength. Today is no different. Labor scholar Kim Moody argues that after decades of capitalist restructuring and change in the composition of the US working class we have arrived at a moment of newfound potential: "We fight on new terrain."[25]

Progressives can't hitch their cart to unions and expect to get where they want to go, however. The sociologist Beverly Silver observes that trade unions occupy an "ambiguous structural position."[26] Unions can be a powerful engine for change, but they ultimately have to look out for and answer to their own membership. To realize the principle that our phones shouldn't be used to perpetuate and obscure coercive and unjust relationships we need a political movement that is bigger than the labor movement, one that re-emphasizes and restores the basic rights of all people in the digital age. The dramatic shifts we've seen over the past decade call for a robust social safety net that includes not only portable benefits, to make equitable flexible work arrangements possible, but also a digital safety net that protects people from surveillance and data theft.[27]

The second principle, that our phones shouldn't be used to mask selfish or immoral behavior, is a bit mushier. It's difficult to organize around social norms and behavioral proscriptions because there may be broad disagreement on what constitutes acceptable behavior, whether we're talking about the behavior of people or institutions. Nonetheless, we see workers and other civil society organizations mobilizing around questions of right and wrong. The clearest example is the recent Google walkouts, which were primarily fueled by a deep sense that the company was pursuing immoral projects and perpetuating unethical behavior. Google employees didn't want to be involved in projects tasked with perfecting drone technology; they didn't want to sign their name to a censored search engine; they refused to look the other way when sexual abusers got a bonus. Tech workers at Amazon demonstrated the same principles when they stood up against their company's selling facial recognition software to police departments; workers at Microsoft and Salesforce joined them when they realized their hard work was being used to separate families trying to cross the US border.

Norms and behaviors might be hard to define, but sometimes they are central. For example, proponents of the Green New Deal advocate a plan in which, over time, countries switch from dirty energy to clean energy. The appeal of this plan is that, in theory, life could go on very much the same after making this switch. Instead of charging our phones and storing our Facebook posts using coal-powered technology we would charge and store using solar or wind energy. Our behavior and norms wouldn't really need to change.

But as environmental justice activists emphasize, our norms and behavior *do* need to change. Even if we switch to green energy—and we absolutely must do this—we still need to take a hard look at the global resource

drain that Western consumers perpetuate. Our consumption patterns, fueled by myriad factors, from planned obsolescence and Apple's refusal of
"right to repair" to the psychological crutch of impulse shopping are not
sustainable. Even as the supply chains that produce the high-tech wonders
of modern capitalism become ever more hidden they are no less destructive
and extractive.

Poor communities, primarily in the Global South, foot the ecological bill for this despoiling. In a high-tech reboot of the colonial division
of labor, corporations destroy farmland, waterways, and mountains so that
wealthy consumers can order a Game of Thrones Monopoly board game on
the way home from work and get it delivered the next day. A "just transition"
must include mechanisms to encourage and enable individuals, households,
and communities to find joy and a sense of purpose through other activities
besides shopping.

Grassroots groups around the country are pursuing this type of just
transition. Cooperation Jackson is a community initiative in Jackson, Mississippi, working toward building a solidarity economy of green, worker-owned
businesses and a broader cooperative network grounded in democratic, eco-
socialist principles. Cooperation Jackson's programs and vision include everything from 3-D printing workshops to calls for mass civil disobedience,
and they emphasize a new kind of environmental justice movement that
bridges the need for sustainable consumption and production with progressive economic policies that address the deep divides in American society.[28]

Ultimately, the fight to change the norms of our smartphone society
is a challenge equal to the struggle against precarity and surveillance, not
least because behaviors and convictions are often individualized, making
them difficult to effectively organize around. This characteristic can encourage us to locate change primarily within ourselves and disconnect our
personal troubles from broader societal issues, or, worse, mistake for a social movement our individual choices to buy fair trade goods or to share
our opinions on Twitter. This doesn't mean that personal decisions and
individual actions aren't important. It simply means that collective action
is much more powerful. As recent initiatives by tech workers show, voicing
our personal concerns about moral and ethical questions collectively in our
schools, communities, and workplaces draws attention to the shared reality
and concerns of our smartphone society; it revives the old union slogan that
an injury to one is an injury to all. Collective action isn't an easy solution.
It's much easier to dash off a line on Facebook or retweet an article that

articulates our feelings. There's nothing wrong with doing these things. They just aren't enough.

The last principle we suggested is that our phones should be a pathway to a true digital commons where our personal data isn't for sale. This principle highlights the deep contradiction between ordinary people's motivations for using their smartphones and those of the tech companies that control the phones' design and functionality. Regular folks are interested in fun, learning, and connection, whereas corporations are interested in marketizing the moments of our lives for a profit. People have historically resisted the commodification of new spheres of life. The push to win control over our phone worlds is part of the long struggle against capitalism's drive to mold society in its own image.

Right now watchdogs and researchers protecting privacy and demanding algorithmic accountability are grappling with the effects of Silicon Valley's drive to transform life into bits and bytes for sale to the highest bidder. But few mainstream groups question the basic notion that it is OK to collect our personal data and sell it. Most groups working today suggest reforms and prescriptions that leave essentially untouched the underlying business model of tech titans whose astronomical profits are derived from mining and selling—commodifying—our personal data. Even the GDPR, which represents a clear step forward and is the most comprehensive data privacy legislation by far, is based on a questionable model of site-by-site individual consent and gives tech companies a great deal of leeway to continue collecting personal data.[29]

There are a few reasons for the reluctance to challenge the collection and sale of personal data. One is practicability. It's easier to achieve concrete reforms by framing the new challenges that have arisen in our smartphone society, such as perpetual surveillance, within past precedents and parameters. Calls to limit the use of facial recognition software or reforms that enable us to see the data that has been collected about us try to rein in the underlying trends of our smartphone society rather than to change them. It's an understandable approach, but also a very limited one. In resigning ourselves to nibbling around the edges of the trends defining the twenty-first century we surrender central parts of our lives to the drives of profit and power.

A second reason we fail to challenge the datafication of life—to press for decommodifying personal data—is that we've been led to believe that restricting the collection and sale of personal data that we generate 24/7 with our hand machines will stifle technological progress and prevent the realization of potential gains from big data. This is an understandable conclusion because it is the one constantly pushed by the Silicon Valley hype machine, but it is incorrect. The avenues of research that could be conducted using nonhuman data are boundless. DeepMind recently used a reinforcement learning algorithm to optimize its electricity use. Other companies use big data on crop yields or supply chains to optimize logistics. Moreover, most of our personal data isn't used for high-minded scientific research. Silicon Valley wants us to think that the no-holds-barred vacuuming of our data exhaust is necessary for the glorious scientific future, but it isn't. It's primarily used to increase engagement with our smartphones by creating addictive apps and to show us ads to get us to buy more junk.

A third countervailing factor against a decommodification approach is that digital justice groups are trying to entice tech companies to willingly accept their agenda. This is a trend we've seen in other social movements, particularly the environmental movement. One of the defining characteristics of neoliberalism is the perception that governments are ineffective and weak, so if we want to get things done we need to partner with corporations who have reach, influence, and an ability to coordinate action quickly. In this ideological climate, groups such as Greenpeace, the Sierra Club, and the World Wildlife Fund have opted to partner with corporations in an attempt to persuade them to voluntarily adopt eco-business practices rather than work for legislation to regulate pollution and extraction.

AI Now, one of the leading groups calling for algorithmic accountability, is taking a similar approach, openly working in collaboration with tech companies and supported financially by the companies creating the problems it aims to solve: Microsoft, Google, DeepMind. Not surprisingly, AI Now's recommendations show the fingerprint of industry, particularly the first recommendation in its 2018 annual report, which advocated for a "sector-specific approach" to regulating algorithms "that does not prioritize the technology, but focuses on its application within a given domain." The National Highway Traffic Safety Administration was proposed as a model. A sector-specific approach is precisely the kind of fix tech companies can live with because it gives them a lot more wiggle room than a centralized body would. Indeed, we've seen this wiggle room in action at the National Highway Traffic Safety Administration for decades, as automotive companies have used their power and influence to ignore, evade, and dilute safety regulations that if followed would have prevented thousands of accidents and deaths.[30]

As in the first Gilded Age, big corporations are attempting to set the boundaries for permissible debate.[31] We shouldn't let them. The tech titans can absorb and implement the critique and reforms mainstream groups are presenting without missing a beat. In doing so they get a stamp of approval to continue to commodify our lives. They get to be portrayed as institutions motivated by the common good. They are not. The devastating critique of black-box algorithms and their detrimental effects made by a growing number of data scientists underscore how important it is that the degree of regulation match the dangers.

Instead of acquiescing to market-friendly reforms restricting how our personal data can be used, we need a clear demand: *Our personal data should not be collected and sold at all.* This regulation would fundamentally change the business models of for-profit tech companies, so it is unlikely that they'd be on board with a project centered on decommodifying our personal data. This is unfortunate, but the stakes are too high to appease all stakeholders; if we want a true digital commons we'll have to agree to disagree with the tech companies.

Some may argue that in today's landscape of territorial control by corporations and governments, the demand for a true digital commons where our personal data isn't collected, bought, and sold is a Utopian demand. It is. Creating a digital commons in which our data is not collected and sold is a special kind of Utopian demand, however, in that it is eminently achievable, particularly in this moment of political flux and discontent. We've long had rules about privacy and about what information should and shouldn't be shared, let alone sold. Creating a digital commons that doesn't rely on commodifying human life is a matter of updating our norms and regulations. Moreover, with the proper regulations, we can still use data, even personal human data, to advance our collective knowledge.

This approach is not without precedent. The United States has strict regulations for the pharmaceutical industry that in large part resulted from public pressure. Drug companies and researchers can't just make a drug and try it out "in the wild" on human subjects. In order to deploy a drug they have to register clinical trials with the FDA, demonstrate efficacy, and show that the drug is safe, that it doesn't have unintended consequences.

We can envision a similar model in developing and using algorithms that rely on human data. Instead of tech companies stealing and scraping data, buying and selling it, in order to "train" their algorithms, which are then tested in the wild, personal data would not be collected and sold at all. If tech companies want to develop algorithms that use human data they would have to conduct clinical trials, under the guidance of a centralized body, using data collected from volunteers for a specific purpose. If the clinical trials were a success and the algorithm proved beneficial and efficacious, with no unintended side effects, the company or research group would be allowed to collect more data and potentially scale the algorithm.[32]

This model wouldn't shut down potential gains from big data and machine learning. It might slow down the implementation of new algorithms that are trained on human data, but that's a good thing. It would make researchers more thoughtful as to how they collect data and design algorithms and may spur a desperately needed conversation about the ecological toll of storing terabytes of data indefinitely. More important, the model would reinforce the idea that people are not products or machines or data, and as such, require special protections—protections the public has long demanded that sectors outside of Silicon Valley respect and obey. Tech companies would like the public to ignore or forget this fact. It's up to us to remind them.

Epilogue

The mass adoption of the automobile marked a new chapter in American life: the automobile was central to the ideas, structures, and struggles that came to define the twentieth century. Today we're once again at the start of a new chapter. Like the automobile, the smartphone has become a defining feature of modern society.

The analogy between the automobile and the smartphone offers more than the reassuring sense of solidity that historical precedent provides. It also offers a positive lesson. As auto workers who staged sit-down strikes in the 1930s showed, we can win power over machines and the corporations who control them. Today, a few powerful corporations dominate the infrastructure of modern social, political, and economic life, but this power isn't divinely ordained. Through a combination of ignorance and resignation we willingly accede to the visions and drives of the tech titans and accept the appropriation of life that undergirds their business model.

It doesn't have to be this way. The new normal of ubiquitous mass surveillance relies on our acquiescence, not to mention our unpaid labor and collectively generated knowledge and infrastructure. Frederick Douglass said, "Power concedes nothing without a demand. It never did and it never will." We have the power to say "Enough!"—to construct a new sociotechnical contract.

Our smartphones are a key site of this struggle. As a cornerstone of the digital frontier and the push to commodify life itself—the datafication and marketization of our interactions, explorations, meanderings, and biorhythms—our pocket computers offer a unique window into contemporary life. The relationships and structures embodied and activated by our phones are a microcosm of modern-day capitalism. Through our smartphones we see how the twenty-first-century working class is being made and the new avenues through which corporate America cements its power and privilege. For better or worse, we also see how ordinary people have incorporated their pocket computers into new strategies to ward off alienation and to find meaning in everyday life. These strategies mark the emergence of new cultural norms and ideas, and more starkly, encapsulate the defining contradiction of the smartphone age: regular folks use their hand machines to socialize, learn, connect, and have fun, while tech companies use them to generate and extract data for a profit.

Our phones are more than a window; they are also proving to be an important tool in the struggle against the status quo. The last decade has been marked by a deep crisis of legitimacy for neoliberal capitalism. Young people in particular have transformed their smartphones into weapons in the struggle for justice and security, repurposing social media as political media. In the process they've catalyzed a new way of doing politics, a digital-analog model that's at the center of a worldwide wave of radical discontent.

Central to this repurposing are the unique capabilities of smartphone technology. Smartphones afford a constant digital pathway, a wherever/whenever connection that facilitates solidarity and collective action in unprecedented ways. Yet we should be wary looking for answers to political questions in smartphone software and hardware. As powerful as our phones have shown themselves to be in connecting people and sharing ideas, there is no technological fix for the challenges of modern life. Our phones are an important tool, but a tool, nonetheless.

The automobile-smartphone analogy also offers a negative lesson: that if we want to challenge the imperatives of Silicon Valley we can't settle for bread-and-butter reforms. Efforts to rein in the auto companies stopped far short of industrial democracy, let alone a mandate for molding the automobile industry to the needs of society rather than molding society to the needs of the auto industry. If we want to truly address the ills of our smartphone society we will have to push beyond minor revisions that nibble around the edges of the tech titans' business model. Wresting control of our pocket computers and our data is essential to actualizing the principles that our phones shouldn't be used to perpetuate and obscure coercive and unjust relationships or to mask bad behavior, and should be a pathway to a true digital commons. After all, our phones are not only a window into a fraught present; they also provide a glimpse into an unsettling future defined by ecological crisis and untrammeled corporate power.

High stakes also mean huge potential gains, however. Claiming smart-phones for people instead of for corporations would establish new norms and expectations of privacy vis-à-vis both the tech behemoths and the state, and more profoundly, would fortify the rapidly collapsing levee holding back the marketization of life itself. A victory this big might seem out of reach, but if our smartphones demonstrate nothing else it is that human ingenuity knows no bounds.

Acknowledgments

Many people helped make this book happen. Karen Narefsky, Remeike Forbes, and Bhaskar Sunkara gave crucial comments on a 2015 essay where I explored some of the key ideas discussed in these pages. Nikil Saval and Jason Farbman both graciously assisted me in finding an agent. My agents, Anthony Arnove and Roisin Davis at Roam Agency, were a huge help as I navigated the process of finding a publisher and finalizing a contract. I'd also like to thank Jessie Kindig (formerly at Roam and now at Verso) for her suggestions on an early draft of my book proposal. Finally, on the publishing end, I owe a great debt of gratitude to my editor, Rakia Clark, who believed in the project from the start and made sure it found a home at Beacon Press. Thanks also to Olivia Bauer, Kate Scott, and Susan Lumenello at Beacon for their helpful comments on the manuscript.

I'd also like to thank Victor Wallis, Suren Moodliar, Itai Vardi, Charles Fisher, and Hareet Sandhu for their useful suggestions and thoughtful conversations. My dear friends Ella Lesatele, Indrani Chatterjee, Myka Tucker Abramson, Alex Gourevitch, and Tal Astrachan provided moral support and lots of laughs. My Canadian friends and mentors Sam and Schuster Gindin have enthusiastically supported the project and brought their wisdom to bear in many discussions about it.

My family has played a central role in helping me write a book that I'm proud to share. Ila and Simi, my vivacious daughters, eagerly drew cover designs and declared topics that *must* find their way into the book. Emilio Sauri and Susana Domingo Amestoy have become a part of my family in the years I've lived in Cambridge. They've been close confidantes throughout the evolution of the project, sending articles and during our Friday-night dinners helping me to flesh out my arguments. My mom, Joan Frisbie, encourages me in all that I do and I often hear her voice as I write. My mother-in-law, Manju Mehta, made me endless cups of tea while I put the finishing touches on the manuscript. My father-in-law, Parmender Mehta, passed away unexpectedly just as I was completing the project, leaving a gaping hole in our family. Papa liked music and mountains much more than technology, but I dedicate this book to him with love and affection.

I'd like to extend a special thanks to Tavish Vaidya. When Tavish first moved to the United States from India he insisted that I get a smartphone, and over the years he has helped me in innumerable ways, including helping me make my website. More important, he readily offers his expertise on topics of computer security and has proved to be an invaluable source of information. Pankaj Mehta, my partner in life, I save for last because at the end of the day he is the person standing beside me, helping me find my voice. In this project, as in all my projects, Pankaj has provided essential emotional, logistical, and intellectual support, from emailing me hundreds of articles with the subject line "Smartphone Book" to explaining how machine learning works. Pankaj often says that he is my biggest fan. The feeling is mutual.

Notes

Introduction

1. Kellaway, "Bliss of Being 396 Miles from My Lost Smartphone."
2. For an interesting behind-the-scenes look at this event and the lead-up to it, see Vogelstein, *Dogfight*.
3. Lynd and Lynd, *Middletown*.
4. Lynd and Lynd, *Middletown*, 137.
5. Lynd and Lynd, *Middletown*, 99, 258.
6. Lynd and Lynd, *Middletown*, 283.
7. Lynd and Lynd, *Middletown*, 153.
8. Lynd and Lynd, *Middletown*, 254.
9. Lynd and Lynd, *Middletown*, 256.
10. Lynd and Lynd, *Middletown*, 61n.
11. Lynd and Lynd, *Middletown*, 33, 35, 74.
12. Leswing, "The Average iPhone Is Unlocked 80 Times Per Day."
13. Robb, "The New Normal."
14. Schmidt, "Always Practise Safe Text."
15. The difference between smartphones and "dumb phones" is not binary. In 2012, most phones used worldwide were "feature phones," providing photo, some web browsing, and app functionality, as well as telephoning and texting. The transition to what we now consider smartphones has really taken hold since then.
16. Purnell, "The Internet Is Filling Up."
17. For a good discussion see Greenfield, *Radical Technologies*, chapter 1.
18. McCarthy, "China Now Boasts More Than 800 Million Internet Users and 98% of Them Are Mobile."
19. Aaron Smith, "Overview of Smartphone Adoption," Pew Research Center, July 11, 2011.
20. Trachtenberg, *The Incorporation of America*, 45–46.
21. For a useful history of the smartphone, see Woyke, *The Smartphone*.
22. Licoppe, "'Connected' Presence," 150.
23. Licoppe, "'Connected' Presence."
24. Ling and Donner, *Mobile Communication*, 92.
25. Turkle, *Reclaiming Conversation*, 13.
26. Eric Pickersgill describes his project on his personal website: https://www.ericpickersgill.com/removed.
27. Schmidt, "New Studies Link Cell Phone Radiation with Cancer."

28. Carr, "How Smartphones Hijack Our Minds." See also Carr, *The Shallows*.

29. See Schüll, "Stuck in the Machine Zone." Her observations about the "machine zone" are developed in Schüll, *Addiction by Design*.

30. Thomas, "Digitally Weary Users Switch to 'Dumb' Phones."

31. For more about Klemens Schillinger's Substitute Phone, see his website: https://klemensschillinger.com/projects/substitute-phone.

32. Rojek, *Presumed Intimacy*, 8–9.

33. Harcourt, *Exposed*.

34. Zuboff, *The Age of Surveillance Capitalism*. Zuboff presents a rich, thoughtful account that should be read by everyone.

35. Price, "How to Break Up with Your Phone."

36. Freed, *Wired Child*, 95. Freed is a practicing psychologist in California. *Wired Child* is his self-published book detailing his clinical cases with teenage technology addiction.

37. Twenge, "Have Smartphones Destroyed a Generation?" For a critique of smartphones and their impact on teenage girls in particular, see Sales, *American Girl*.

38. Wajcman, *Pressed for Time*, 150.

39. Ling, *New Tech, New Ties*, 164–68.

40. Thompson, *Smarter Than You Think*.

41. Willingham, "Smartphones Don't Make Us Dumb."

42. American Academy of Pediatrics, "American Academy of Pediatrics Announces New Recommendations for Children's Media Use," October, 21, 2018.

43. Friedman, "The Big Myth About Teenage Anxiety."

44. Przybylski and Weinstein, "A Large Scale Test of the Goldilocks Hypothesis."

45. Margolis, "A Crash Course in Technology for Ageing and Infirm Bodies."

46. For more information on these technologies, see the Sesame Enable website, https://sesame-enable.com.

47. Pew Research Center, "Internet/Broadband Fact Sheet," June 12, 2019, www.pewinternet.org/fact-sheet/internet-broadband. (NB: Pew updates its fact sheets on a regular basis.)

48. Wacjman, *Pressed for Time*, 21.

49. Spigel, *Make Room for TV*, 3, 46.

50. Lears, *Rebirth of a Nation*, 8.

51. Benjamin, "The Work of Art in the Age of Mechanical Reproduction," 91.

52. Postman, *The Disappearance of Childhood*.

53. Cited in Spigel, *Make Room for TV*, 54.

54. Putnam, *Bowling Alone*.

55. For a useful discussion of technological determinism, see the introductory essay in MacKenzie and Wajcman, *The Social Shaping of Technology*.

56. Zucman, "Global Wealth Inequality."

57. Temin, *The Vanishing Middle Class*.

58. Federal Reserve Bank of New York, "Household Debt Jumps as 2017 Marks the Fifth Consecutive Year of Positive Annual Growth Since Post-Recession Deleveraging," press release, February 13, 2018.

59. Board of Governors of the Federal Reserve System, "Report on the Economic Well-Being of US Households in 2017," May 2018.

60. Pew Research Center, "Public Trust in Government: 1958–2017," *US Politics and Policy*, December 14, 2017.

61. Kavanagh and Rich, *Truth Decay*.

62. See, for example, Ferguson, Jorgensen, and Chen, "Industrial Structure and Party Competition in the Age of Hunger Games."

63. Winner, "Do Artifacts Have Politics?," 122.

64. Braverman, *Labor and Monopoly Capital*, 12.

Chapter One: New Divides (or, Old Divides Made New)

1. Tanya Marshall and the names associated with this story are pseudonyms, but the facts of the story are true and were recounted to me personally by Tanya.

2. Levin, "Officer Punched Oscar Grant and Lied About Facts in 2009 killing, Records Show."

3. Babwin and Tarm, "Officer Convicted of Murder in Slaying of Laquan McDonald."

4. Hassan, "Family of Jordan Edwards Says 15 Years Is Not Enough for Officer Who Murdered Him."

5. Dewan and Oppel, "In Tamir Rice Case, Many Errors by Cleveland Police."

6. "The Counted" was a project by the *Guardian* newspaper to create a comprehensive list of people killed by the police and other law enforcement agencies in the United States in 2015 and 2016. The project was undertaken because the US government doesn't keep a comprehensive list of how many people are killed by law enforcement, and police departments are not required to submit detailed reports on the use of force against civilians. See "About the Project: What Is The Counted?," https://www.theguardian.com/us-news/ng-interactive/2015/jun/01/about-the-counted.

7. Taylor, *From #BlackLivesMatter to Black Liberation*, 3.

8. "Meet Abdullah Muflahi: He Filmed Alton Sterling's Shooting & Was Then Detained by Baton Rouge Police," *Democracy Now!*, July 13, 2016.

9. Chappell, "Philando Castile Is Remembered by St. Paul Public Schools."

10. Pearce, Hennessy-Fiske, and Evans, "As Police Shootings Continue."

11. Laughland and Holpuch, "Walter Scott."

12. Pearce, Hennessy-Fiske, and Evans, "As Police Shootings Continue."

13. Smartphone users in Michigan can download both an iOS and Android version of the Mobile Justice app. For more information, see ACLU of Michigan, "Mobile Justice MI Program," https://www.aclumich.org/en/campaigns/mobile-justice-mi-project.

14. Torre, "Indigenous Australians Use Tech to Expose Police Abuse."

15. Jasper, "He Took Video Outside an ICE Raid in Sanford. Now He's Charged with Communicating Threats."

16. For a comprehensive review of existing studies on the use of body-worn cameras by US police departments, see Lum et al., "Research on Body-Worn Cameras," 18.

17. Levin, "Police Shot a Pregnant California Teen."

18. Rudder, *Dataclysm*, 9.

19. Cacioppo et al., "Marital Satisfaction and Break-Ups."

20. Winkie, "Why Men Are Paying Women for Accurate Critiques of Their Dicks."

21. These findings are from the 1977 General Social Survey. For a comprehensive discussion, see Cherlin, *Labor's Love Lost*, 15.

22. Cherlin, *Labor's Love Lost*, 138. See also Cherlin, *The Marriage-Go-Round*.

23. Sales, "Tinder and the Dawn of the 'Dating Apocalypse.'"

24. Hodgson, "Inside the Secretive World of Stalking Apps."

25. Cloos and Turkewitz, "Hundreds of Nude Photos Jolt Colorado School."

26. The bill, H.R. 1761, Protecting Against Child Exploitation Act of 2017, updated existing legislation. For more, see Burns, "House Passes Bill That Could Have Teens Facing 15 Years for Trying to Sext."

27. For an in-depth discussion, see Hasinoff, *Sexting Panic*.

28. Rosin, "Why Kids Sext."

29. Orenstein, *Girls & Sex*.

30. Hasinoff, *Sexting Panic*, 109, 97.

31. Aschoff, "The Smartphone Society," 36.

32. Gereffi, "Beyond the Producer-Driven/Buyer-Driven Dichotomy," 3.

33. Lee and Gereffi, "The Co-Evolution of Concentration in Mobile Phone Global Value Chains."

34. See Abraham, *The Elements of Power*.

35. Jameson, "The End of Temporality," 4, 700.

36. *Digital Dividends*, World Development Report (Washington, DC: World Bank Group, 2016).

37. Pew Charitable Trust, "24 Million Americans Still Lack Broadband Connectivity," fact sheet, July 24, 2018.

38. Upgrade Cambridge, "Percent of Cambridge Households with Broadband Subscription by Household Income 2013–2016," graph, https://upgradecambridge.org/digital-equity-in-cambridge.

39. Strain, Moore, and Gambhir, "AT&T's Digital Divide in California."

40. "AT&T's Digital Redlining," National Digital Inclusion Alliance, March 2017.

41. For an illuminating look at these digital divides, see Maria Smith's video project *Dividing Lines*, https://www.dividinglines.org.

42. Hochschild, *The Time Bind*.

43. Lareau, *Unequal Childhoods*.

44. Board of Governors of the Federal Reserve System, "Report on the Economic Well-Being of U.S. Households in 2017–May 2018," May 2018; US Bureau of Labor Statistics, "Contingent and Alternative Employment Arrangements News Release," June 7, 2018.

45. Weil, *The Fissured Workplace*, 87–88, 7, 44.

46. "TNCs & Congestion," draft report (San Francisco: San Francisco County Transportation Authority, October 2018).

47. Soper, "Smartphone Shopping."

48. Sainato, "'We Are Not Robots.'"

49. Bruder, *Nomadland*, 62.

50. Lee et al., "Working with Machines."

51. O'Connor, "When Your Boss Is an Algorithm."

52. JC, "Ridester's 2018 Independent Driver Earnings Survey," Ridester, March 29, 2019, www.ridester.com/2018-survey.

53. For a discussion, see Taplin, *Move Fast and Break Things*.

54. Friedman, *The World Is Flat*.

55. Waddell, "Why Bosses Can Track Their Employees 24/7."

56. Levin, "Sexual Harassment and the Sharing Economy."

57. See Laufer, "Social Accountability and Corporate Greenwashing."

58. Amazon.com, "2017 Amazon Holiday Commercial," video, https://www.youtube.com/watch?v=OITWgx8K6Ko.

59. Deleuze, "Postscript on the Societies of Control," 6.

Chapter 2: New Titans

1. General Service Studios, "A Trip Through the River Rouge Plant," film, n.d., https://archive.org/details/TripThroughTheRiverRougePlant.

2. Lears, *Rebirth of a Nation*, 51, 57, 61.

3. Brands, *American Colossus*, 7–8.

4. Roose, "Forget Washington."

5. Moazed and Johnson, *Modern Monopolies*, 5.

6. Ghose, *Tap*, 18.

7. For more on this subject, see Levine, *Surveillance Valley*. See also Mazzucato, *The Entrepreneurial State*.

8. Matt Day, "Amazon Profits Are Floating High on the Cloud," *Fortune*, April 25, 2019.

9. Helft, "Meet the Man Who Derailed Facebook's Plan to Provide Free Internet in India."

10. Facebook has quietly ended Free Basics in a number of countries, including Myanmar.

11. Hollister, "Facebook Is Officially Building an Internet Satellite: Athena."

12. Emily Bary, "Apple Worth $1 Trillion for First Time in 2019 after Detailing New iPhones, Streaming Pricing," *MarketWatch*, September 14, 2019; Courtney Dentch, "Microsoft Reclaims $1 Trillion Mark as Stock Climbs to Record," *Bloomberg*, June 7, 2019; McChesney, *Digital Disconnect*, 112.

13. Woetzel et al., "China's Digital Economy."

14. Autor et al., "Concentrating on the Fall of the Labor Share," 180–185.

15. For an excellent analysis, see Gertner, *The Idea Factory*.

16. See, for example, Panitch and Gindin, *The Making of Global Capitalism*.

17. For an excellent historical discussion, see Khan, "Amazon's Antitrust Paradox."

18. Mullins, Winkler, and Kendall, "Inside the U.S. Antitrust Probe of Google."

19. Thiel, *Zero to One*.

20. See, for example, Moazed and Johnson, *Modern Monopolies*.

21. For the full text of Jeff Bezos's letter see http://media.corporate-ir.net/media_files/irol/97/97664/reports/Shareholderletter97.pdf.

22. Solon and Siddiqui, "Forget Wall Street."

23. Lee, "Google Is Losing Allies."

24. Bill Galston and Bill Kristol, "Big Tech: Public Discourse and Policy," New Center, https://cloudfront.newcenter.org/wp-content/uploads/2018/11/26062745/Big-Tech-Policy-Paper2.pdf.

25. Yang, "Apple Investigates Illegal Student Labour at Watch Assembly Plant."

26. Kantor and Streitfeld, "Inside Amazon."

27. See, for example, a letter sent to Google CEO Sundar Pichai by Google employees on December 5, 2018, https://medium.com/@GoogleWalkout/invisible-no-longer-googles-shadow-workforce-speaks-up-9ea04b7bcc41.

28. Levin, "Google Accused of 'Extreme' Gender Pay Discrimination."

29. For the full text of the article see Lecher, "An Internal Google Email."

30. Tiku, "Google Deliberately Confuses Its Employees, Fed Says."

31. Angwin and Parris, "Facebook Lets Advertisers Exclude Users by Race"; Angwin, Tobin, and Varner, "Facebook (Still) Letting Housing Advertisers Exclude Users by Race"; Tobin, "HUD Sues Facebook over Housing Discrimination."

32. See Celia Cucalon, "The Silicon Valley Land Grab," Community Legal Services, https://clsepa.org/media-great-silicon-valley-land-grab. Community Legal Services is a

nonprofit legal aid service offering support for families in East Palo Alto, California, on issues of immigration, housing, and worker and consumer rights.

33. For an overview of Silicon Valley Rising and resources about the role of tech companies in shaping the region, see https://siliconvalleyrising.org.

34. Simmons, "How Apple—and the Rest of Silicon Valley—Avoids the Tax Man."

35. Zucman, "How Corporations and the Wealthy Avoid Taxes (and How to Stop Them)."

36. Stampler, "Amazon Will Pay a Whopping $0 in Federal Taxes on $11.2 Billion Profits."

37. Gottfried and Shearer, "News Use Across Social Media Platforms 2017."

38. Mozur, "A Genocide Incited on Facebook."

39. For an in-depth discussion of this problem, see McChesney, *Digital Disconnect.*

40. This is not just a problem for newspapers. Print magazines and book publishers are also in crisis. Publishing powerhouse Condé Nast lost $120 million in 2017. As Amazon has tightened its stranglehold on book publishing, authors have seen their livelihoods destroyed. According to the Authors Guild, the median pay of a full-time writer today is $20,300. See "Six Takeaways from the Author's Guild 2018 Author Income Survey," www.authorsguild.org/industry-advocacy/six-takeaways-from-the-authors-guild-2018-authors-income-survey.

41. Habermas, *The Structural Transformation of the Public Sphere*, cited in McChesney, *Digital Disconnect*, 66.

42. Bell, "Facebook Is Eating the World."

43. For a detailed analysis of this transition see Zuboff, *Surveillance Capitalism,* chapter 3.

44. The *Financial Times* has followed these cases closely. See, for example, Toplensky, "EU Fines Google €2.4bn over Abuse of Search Dominance"; Waters, Toplensky, and Ram, "Brussels' €2.4bn Fine Could Lead to Damages Cases and Probes in Other Areas of Search;" Barker and Khan, "EU Fines Google Record €4.3bn over Android."

45. For a good synopsis of Pariser's ideas, see *Beware Online Filter Bubbles*, video of TED Talk, https://www.ted.com/talks/eli_pariser_beware_online_filter_bubbles?language=en; see also Pariser, *The Filter Bubble.*

46. Zuckerberg himself used to refer to Facebook as a "social utility," but in recent years he has eschewed this terminology, possibly because the implications of Facebook's being a utility are far afield from his vision for the company.

47. For more on neoliberalism, see Cahill and Konings, *Neoliberalism.*

48. Semuels, "How Amazon Helped Kill a Seattle Tax on Business."

49. Goodman, "Amazon Pulls Out of Planned New York City Headquarters."

50. Mullins and Nicas, "Paying Professors: Inside Google's Academic Influence Campaign."

51. Foroohar, "Release Big Tech's Grip on Power."

52. Taplin, "Google's Disturbing Influence over Think Tanks."

53. Marriage and Waters, "Alphabet Set to Face Down Shareholder Dissent."

54. Dobson, "Inside the Verdura Resort in Sicily, Home to Google's Top Secret Summer Camp."

55. Rushe, "Scholar Says Google Criticism Cost Him Job."

56. Foroohar, "Big Tech's Grip."

57. Lears, *Rebirth of a Nation*, 298.

Chapter 3: New Frontier

1. Bennett, "Kim Kardashian Just Wants to Be Seen."

2. Kardashian, *Selfish.*

3. Gary Vaynerchuk, "Do What You Love (No Excuses!)," Ted Talk, Web 2.0 Expo, September 2008.

4. Burgess, Marwick, and Poell, *SAGE Handbook of Social Media*, introduction.

5. Rojek, *Presumed Intimacy*, 135.

6. Ling, *New Tech, New Ties*, 43.

7. Lewis and Jacobs, "How Business Is Capitalising on the Millennial Instagram Obsession."

8. Lewis and Jacobs, "How Business Is Capitalising on the Millennial Instagram Obsession."

9. The original paper on the experience economy is Pine and Gilmore, "Welcome to the Experience Economy."

10. For more on habitus, see Bourdieu, *Distinction.*

11. Munro, *A Wilderness Station*, 99–116.

12. Goffman, *The Presentation of Self in Everyday Life.*

13. Portions of this section appeared in Aschoff, "The Smartphone Society," 38.

14. Chopik, "The Benefits of Social Technology Use Among Older Adults Are Mediated by Reduced Loneliness."

15. Wilson and Yochim, "Pinning Happiness," 234–36.

16. Lucero, "Safe Spaces in Online Places."

17. Rojek, *Celebrity*, 52–53, 90–91.

18. Pew Research Center, "Majority of Users Say It Would *Not* Be Hard to Give Up Social Media," *Social Media Use in 2018* (February 27, 2018).

19. Sullivan, "I Used to Be a Human Being."

20. Sullivan, "I Used to Be a Human Being."

21. Harcourt, *Exposed*, 50.

22. Gregory, "This Startup Wants to Neutralize Your Phone."

23. O'Neil, "My Candid Conversations with Extremely Online Folks Who Suffer from Internet Broken Brain."

24. Luckerson, "The Rise of the Like Economy."

25. See Vincent, "Former Facebook Exec Says Social Media Is Ripping Apart Society"; Chamath Palihapitiya, "Chamath Palihapitiya, Founder and CEO Social Capital, on Money as an Instrument of Change," video of talk at Stanford Graduate School of Business, uploaded November 13, 2017, is available on YouTube, https://www.youtube .com/watch?v=PMotykwoSIk&feature=youtu.be&t=21m21s.

26. Lewis, "'Our Minds Can Be Hijacked.'"

27. Nicas, "How YouTube Drives People to the Internet's Darkest Corners."

28. Tufekci, "YouTube, the Great Radicalizer."

29. Lanier, *Ten Arguments.*

30. Siva Vaidhyanathan said this during an interview with Amy Goodman and Juan González on *Democracy Now!*, August 1, 2018.

31. For a discussion, see Federici, *Caliban and the Witch*, 152. See also Hillis, *Online a Lot of the Time*, 19.

32. Gabler, *Life: The Movie*, 25.

33. Gabler, *Life: The Movie*, 29.

34. Haraway, "A Cyborg Manifesto."

35. Eva Illouz, *Cold Intimacies*, 91.

36. Goffman, *Presentation of Self in Everyday Life*, 36.

37. Lanier, *Ten Arguments*, 6.

38. Bernasek and Morgan, *All You Can Pay*.

39. Levine, *Surveillance Valley*, 159–60.

40. Nakashima, "AP Exclusive." Reports on user data that Google collects have proliferated. For a partial list, see, for example, Curran, "Are You Ready? Here Is All the Data Facebook and Google Have On You"; Macauley, "What Google Knows About You."

41. Viljoen, "Facebook's Surveillance."

42. See, for example, Fried, "What Facebook Knows About You."

43. Vigneri et al., "Taming the Android AppStore."

44. Schneier, *Data and Goliath*, 17.

45. For a detailed discussion, see Moore, *Capitalism in the Web of Life*.

46. Moore, *Capitalism in the Web of Life*, 54.

47. Moore, *Capitalism in the Web of Life*, 63, 69, 101.

48. Moore, *Capitalism in the Web of Life*, states, "We are dealing with a new era: the end of Cheap Nature" (108).

49. Federici, *Caliban and the Witch*, 97

50. James, *Sex, Race, and Class*, 45.

51. James, *Sex, Race, and Class*, 50.

52. Machine learning (often conflated with artificial intelligence) refers to a field of study whose goal is to learn statistical models directly from large datasets. Machine learning also refers to the software and algorithms that implement these statistical models and make predictions on new data.

53. Mayer-Schönberger and Cukier, *Big Data*, 93.

54. Levine, *Surveillance Valley*, 153.

55. Facebook's financials can be found in its 2018 10-K filing for the Securities and Exchange Commission: https://www.sec.gov/Archives/edgar/data/1326801/000132680119000009/fb-12312018x10k.htm.

56. Molla, "Google Leads the World in Digital and Mobile Ad Revenue."

57. Melendez and Pasternack, "Here Are the Data Brokers Quietly Buying and Selling Your Personal Information."

58. For an excellent discussion, see Harcourt, *Exposed*, 198–208.

59. Wu, *The Attention Merchants*.

60. Harcourt, *Exposed*, 202.

61. Dixon and Gellman, "The Scoring of America."

62. Safdar, "On Hold for 45 Minutes?"

63. Dixon and Gellman, "The Scoring of America."

64. Ghose, *Tap*, 93.

65. Murgia and Ralph, "Facebook Blocks Admiral."

66. Murgia and Ralph, "Facebook Blocks Admiral."

67. Ralph, "Insurance and the Big Data Technology Revolution."

68. Trachtenberg, *Incorporation of America*, 37.

69. Cited in McChesney, *Digital Disconnect*, 101.

70. McChesney, *Digital Disconnect*, 162.

71. Levine, *Surveillance Valley*, 33, 75.

72. Sledge, "CIA's Gus Hunt on Big Data."

73. Schneier, *Data and Goliath*, 5.

74. This doesn't mean that the CIA can break encryption. It means that the agency has developed tools to access devices themselves so they can gather information *before* it is encrypted. For a useful discussion, see Gillis, "Wikileaks Only Told You Half the Story."

75. Schneier, *Data and Goliath*, 25.

76. Hu, *A Prehistory of the Cloud*, 66, 124, 11.

77. Belkhir and Elmeligi, "Assessing ICT Global Emissions Footprint." Google toyed with the idea of a more eco-friendly, modular smartphone with replaceable parts (Project Ara), but it abandoned the idea in 2016. Statt, "Google Confirms the End of Its Module Project Ara Smartphone."

78. Mayer-Schönberger and Cukier, *Big Data*, 99.

79. Schneier, *Data and Goliath*, 56.

80. See, for example, Singer, "How Google Took Over the Classroom."

81. For more on Minutiae, see the App Store Preview, https://itunes.apple.com/us/app/mi-nu-ti-ae/id1192323032?mt=8.

82. Federici, *Caliban and the Witch*, 75.

Chapter 4: New Politics

1. Tweeted May 8, 2013.

2. Tweeted November 11, 2017.

3. Tweeted May 5, 2016.

4. Kuchler, "US Midterms."

5. Lewis, "History Makers."

6. Mehul Srivastava, "Seven-Year-Old Bana Al-Abed, the 'Face of Aleppo.'"

7. Witt, "How the Survivors of Parkland Began the Never Again Movement."

8. Ling and Donner, *Mobile Communication*, 114–15.

9. Jaffe, *Necessary Trouble*, 220.

10. Abu-Lughod, *Before European Hegemony*, 369.

11. Tufekci, *Twitter and Tear Gas*, 111.

12. For details of the story, see Tufekci, *Twitter and Tear Gas*, 53–61.

13. Figueroa O'Reilly, "Trump Blocked Me on Twitter. Not Any More."

14. The Knight First Amendment Institute has details of the ruling, legal documents regarding the case, and a video recording of the oral argument; see "Knight Institute v. Trump—Lawsuit Challenging President Trump's Blocking of Critics on Twitter," https://knightcolumbia.org/content/knight-institute-v-trump-lawsuit-challenging-president-trumps-blocking-critics-twitter.

15. Kazmin, "'I Am a Troll' by Swati Chaturvedi."

16. Information about NationBuilder software and campaigns can be found on the company's website, https://nationbuilder.com.

17. Ling and Donner, *Mobile Communication*, 117.

18. Yang, "Beijing Now Able to Flag Weibo Posts as 'Rumour.'"

19. Wang, "China's Chilling 'Social Credit' Blacklist."

20. Feng, "Beijing Widens Control of Wildly Popular Short-Video Apps."

21. Barnes, "Saudi Arabia Prosecutor Says People Who Post Satire on Social Media Can Be Jailed."

22. Jon Lee Anderson, who wrote a widely read biography of Che Guevara, wrote a thoughtful piece on ZunZuneo for the *New Yorker*: "The Dangerous Absurdity of the Secret 'Cuban Twitter,'" April 4, 2014.

23. Scahill, *The Assassination Complex*, 9.

24. Snowden, "Foreword: Elected By Circumstance," xvii.

25. Cagle, "Facebook, Instagram, and Twitter Provided Data Access." For more on the CIA's interest in social media mining and surveillance, see Fang, "The CIA Is Investing in Firms."

26. See Geofeedia, "Baltimore County Police Department and Geofeedia Partner to Protect the Public During Freddie Gray Riots," www.aclunc.org/docs/20161011 _geofeedia_baltimore_case_study.pdf.

27. Leetaru, "Geofeedia Is Just the Tip of the Iceberg."

28. The City of Memphis denied any wrongdoing, but as a part of the lawsuit brought by the ACLU of Tennessee, *ACLU of Tennessee, Inc. vs. the City of Memphis*, it unsealed documents related to the case. See "City Unseals Documents in Ongoing Lawsuit," City of Memphis website, https://www.memphistn.gov/news/what_s_new/city _unseals_documents_in_ongoing_lawsuit.

29. Freedom on the Net 2016, "Silencing the Messenger."

30. Neoliberalism itself was a set of ideas and practices that crystallized in response to a different crisis—the triple crisis (political, economic, social) of the 1970s. Expenditure overruns and stalled productivity combined with skyrocketing inflation and erratic financial flows to generate a situation of severe economic turmoil. The effects of this turmoil on working people, amid a broader dissatisfaction with the status quo, led to mass strikes and social movement organizing around issues of racism, sexism, colonialism, and consumer rights fomenting a widespread and disruptive social crisis. By the end of the 1970s, the economic and social crises coalesced into a severe political crisis in the US, encapsulated in Carter's 1979 "crisis of confidence" speech and an emergent bipartisan elite consensus to abandon Keynesianism. See Aschoff, "America's Tipping Point?," 305–25.

31. King, "#blacklivesmatter."

32. King, "#blacklivesmatter."

33. Meyerson, "The Founders of Black Lives Matter.'"

34. King, "How Three Friends."

35. For a detailed review of these policies, see "Policing and Profit," Criminal Procedure, *Harvard Law Review*, June 6, 2019.

36. For a deeper discussion of the Movement for Black Lives, see Taylor, *From #BlackLivesMatter*; Ransby, *Making All Black Lives Matter*.

37. Taylor, *#BlackLivesMatter*, 162.

38. Taylor, *#BlackLivesMatter*, 19.

39. McCarthy, "Americans Maintain a Positive View of Bernie Sanders."

40. *Dakota Access Pipeline Company Attacks Native American Protesters with Dogs and Pepper Spray*, video, *Democracy Now!*, September 4, 2016.

41. Witt, "The Optimistic Activists for a Green New Deal."

42. Solnit, "Standing Rock Inspired Ocasio-Cortez to Run."

43. See also Aschoff, "The Glory Days Are Over."

44. Mudde, "Don't Be Fooled."

45. See Aschoff, *The New Prophets of Capital*, chapter 2.

46. The Heritage Foundation has posted Santelli's rant to YouTube: https://www .youtube.com/watch?v=zp-Jw-5Kx8k.

47. Cited in Graham, *Cities Under Siege*, xx.

48. Wilson, "Who Are the Proud Boys?"

49. Cited in Ferguson et al., "Industrial Structure," 113.

50. Romero, "US Terror Attacks Are Increasingly Motivated by Right-Wing Views."

51. Cubarrubia, "Beyonce Calls Herself a 'Modern-Day Feminist'"; Sandberg, *Lean In*; for a critique see Aschoff, *New Prophets*, chapter 1.

52. Taylor, *#BlackLivesMatter*, 165.

53. Kantor and Twohey, "Harvey Weinstein Paid Off Sexual Harassment Accusers for Decades."

54. Charles, "As a Black, Gay Woman I Have to Be Selective in My Outrage."

55. Collins and Cox, "Jenna Abrams, Russia's Clown Troll Princess, Duped the Mainstream Media and the World."

56. Sunstein, *#Republic*.

57. Morozov, "From Slacktivism to Activism."

58. "That's the Whole Point of Apartheid, Jerry," *Comedians in Cars Getting Coffee*, season 6, episode 5.

59. Fraser, "From Progressive Neoliberalism to Trump and Beyond."

60. Branch and Mampilly, *Africa Uprising*, 83.

61. Jameson, *Archaeologies of the Future*.

62. Bourdieu, "A Reasoned Utopia and Economic Fatalism."

Chapter 5: New Spirit

1. Agnew, "Mamoudou Gassama."

2. For a discussion of storytelling in politics, see Polletta, *It Was Like a Fever*.

3. This discussion draws from and builds on my first book. See Aschoff, *The New Prophets of Capital*, 1–3.

4. Boltanski and Chiapello, *The New Spirit of Capitalism*, 11.

5. Wu, *Attention Merchants*.

6. Blythe, *Austerity*, 44.

7. Trachtenberg, *Incorporation of America*, 14.

8. Ullman, *Close to the Machine*, 34.

9. Krugman, "Why Most Economists' Predictions Are Wrong."

10. Graeber, "Of Flying Cars and the Declining Rate of Profit."

11. Galloway, *The Four*.

12. Bradshaw, "Google Chief Touts Utopian Ambitions."

13. Cited in Sundararajan, *The Sharing Economy*, 7.

14. Bowles, "Clooneys to Attend a VIP Fundraiser for Clinton—and Sanders Fans Are Outraged."

15. Weiss, *The American Myth of Success*, 9.

16. For a good critique of platforms, see Srnicek, *Platform Capitalism*.

17. Sundararajan, *Sharing Economy*, 124.

18. Moazed and Johnson, *Modern Monopolies*, 62.

19. Brynjolfsson and McAfee, *The Second Machine Age*, 106.

20. Anderson, "The End of Theory."

21. Harris, "Inside the First Church of Artificial Intelligence."

22. O'Gieblyn, "Ghost in the Cloud."

23. Ullman, *Life in Code*, 295.

24. Losse, *The Boy Kings*, 201–2.

25. See John Perry Barlow, "A Declaration of the Independence of Cyberspace," Electronic Frontier Foundation website, https://www.eff.org/cyberspace-independence.

26. On Fordlandia, see Grandin, *Fordlandia*.

27. Popper, "A Cryptocurrency Millionaire Wants to Build a Utopia in Nevada."

28. Trachtenberg, *Incorporation of America*, 39.

29. For a crisp discussion of technology, fiction, and cyborg visions, see McCracken, "Cyborg Fictions."

30. Fowler's letter appears on her blog and is entitled "Reflecting on One Very, Very Strange Year at Uber," February 19, 2017. Fowler was named Person of the Year by the *Financial Times*. See Hook, "The Software Engineer Who Lifted the Lid on Sexual Harassment at Uber."

31. Hook and Kuchler, "Uber's Turmoil Compounded by David Bonderman's Sexist Quip."

32. Newcomer, "In Video Uber CEO Argues with Driver over Falling Fares."

33. More information can be found at Gilliard's website: https://hypervisible.com.

34. O'Neil, *Weapons of Math Destruction*, 12.

35. Whittaker et al., *AI Now Report: 2018*, 18.

36. Harari, *21 Lessons for the 21st Century*.

37. Bowles, "Tech CEOs Are in Love with Their Principal Doomsayer."

38. Brooks, "The Seven Deadly Sins of AI Prediction."

39. Atkinson, "Relax. Robots Won't Cause Us to Run Out of Jobs."

40. Noble, *Forces of Production*, xiii.

41. Noble, *Forces of Production*, xiii-xv.

42. Roy, "Why the Constant Struggle to Manage Our Inboxes Is About More Than Just Work."

43. Bowles, "Silicon Valley Nannies Are Phone Police for Kids."

44. Winner, "Do Artifacts Have Politics?"

45. Thomas and Lim, "On Maids, Mobile Phones, and Social Capital."

46. Lanier, *Ten Arguments for Deleting Your Social Media Accounts Right Now*, 106.

47. Posner has written about this concept a number of times; see, for example, Posner and E. Glen Weyl, "Want Our Personal Data? Pay For It."

48. Schneier, *Data and Goliath*, 58.

49. Stallman, "Talking to the Mailman."

50. National Research Council, *Funding a Revolution: Government Support for Computing Research* (Washington, DC: National Academies Press, 1999).

51. Mazzucato, *Entrepreneurial State*, 94.

52. For a deeper analysis of the digital commons, see Taylor, *The People's Platform*.

53. Brands, *American Colossus*, 490.

54. For more on free software, see https://www.gnu.org/philosophy/free-sw.html.

Chapter 6: New Map

1. See Jameson, *Postmodernism*, chapter 1. For a discussion of mental maps, see also Lynch, *The Image of the City*; Sauri, "Cognitive Mapping."

2. Milgram and Blass, *The Individual in a Social World*. For an interesting discussion, see also Yan Leng, Santistevan, and Pentland, "Familiar Stranger."

3. Rivoli, "New York City Will Propose Minimum Wage for Uber, App Drivers."

4. Desai, "How to Beat Uber."

5. For information on ongoing campaigns see the website of the New York Taxi Workers Alliance, http://www.nytwa.org. See also Fitzsimmons, "Uber Hit with Cap as New York City Takes Lead in Crackdown." Uber responded to the cap by suing the city in February 2019 to overturn the cap. The case was ongoing when this book went to

press. For more, see Hawkins, "Uber Sues to Overturn New York City's Cap on New Ride-Hail Drivers."

6. Roose, "After Uproar, Instacart Backs Off Controversial Tipping Policy."

7. Siegel, "DoorDash to Change Its Controversial Tipping Policy After Outcry."

8. "Tech's Invisible Workforce," Silicon Valley Rising, March 2016.

9. Wagner, "Facebook to Raise Pay for Thousands of Contract Workers, Including Content Moderators."

10. For the full petition text, see Tech Workers Coalition, "To: Google, Amazon, Microsoft, IBM: Tech Should Not Be in the Business of War," https://www.coworker .org/petitions/tech-should-not-be-in-the-business-of-war.

11. Kim (pseudonym), "Tech Workers Versus the Pentagon," interview by Ben Tarnoff, *Jacobin*, June 6, 2018.

12. For an in-depth report, see Colvin, "The Growing Use of Mandatory Arbitration."

13. American Civil Liberties Union, Complaint for Injunctive Relief, United States District Court Western District of New York, Case No. 18-cv-1488, December 21, 2018.

14. See, for example, Cagle and Ozer, "Amazon Teams Up with Government to Deploy Dangerous New Facial Recognition Technology."

15. Gosselin, "IBM Accused of Violating Federal Anti-Age Discrimination Law."

16. Merrill and Tobin, "Facebook Moves to Block Ad Transparency Tools."

17. ILSR has a range of resources concerning community broadband; see the Institute for Local Self-Reliance website, https://ilsr.org/broadband.

18. ILSR's Stacy Mitchell has been a particularly vocal critic of Amazon. See, for example, Mitchell, "Amazon Doesn't Just Want to Dominate the Market."

19. The best place for information on the EU's General Data Protection Regulation is https://eugdpr.org.

20. Data & Society is a nonprofit research institute "focused on the social and cultural issues arising from data-centric technological development." For more, see Caplan et al., "Algorithmic Accountability: A Primer," prepared for the Congressional Progressive Caucus, Tech Algorithm Briefing, "How Algorithms Perpetuate Racial Bias and Inequality," Data & Society, April 18, 2018.

21. Diakopoulos and Friedler, "How to Hold Algorithms Accountable."

22. The term "Green New Deal" has been around for more than a decade, but its recent usage refers to a pair of resolutions submitted to Congress by Representative Alexandria Ocasio-Cortez and Senator Ed Markey and to the broader social movement supporting the ideas articulated in the proposals. The Sunrise Movement, a major supporter, has information on its website, https://www.sunrisemovement.org/gnd.

23. Pollin, "De-Growth vs. A Green New Deal."

24. For a discussion, see Silver, *Forces of Labor*.

25. Moody, *On New Terrain*.

26. Silver, "Workers of the World."

27. See, for example, Scholz, "Platform Cooperativism vs. the Sharing Economy."

28. Akuno and Nangwaya, *Jackson Rising*.

29. Stallman, "A Radical Proposal to Keep Your Personal Data Safe."

30. Whitaker et al., *AI Report 2018*.

31. Lears, *Rebirth of a Nation*, 87–88.

32. This idea was developed in conversation with Pankaj Mehta, a theoretical physicist who works on problems in machine learning.

Works Cited

Abraham, David S. *The Elements of Power: Gadgets, Guns, and the Struggle for a Sustainable Future in the Rare Metal Age.* New Haven, CT: Yale University Press, 2015.

Abu-Lughod, Janet L. *Before European Hegemony: The World System A.D. 1250–1350.* New York: Oxford University Press, 1989.

Agnew, Harriet. "Mamoudou Gassama: What the 'Spider-Man' of Paris Did Next." *Financial Times,* January 25, 2019.

Akuno, Kali, and Ajamu Nangwaya. *Jackson Rising: The Struggle for Economic Democracy and Black Self-Determination in Jackson, Mississippi.* Quebec: Daraja Press, 2017.

Anderson, Chris. "The End of Theory: The Data Deluge Makes the Scientific Method Obsolete." *Wired,* June 23, 2008.

Anderson, Janna. "Future of the Internet IV." Pew Research Center, February 19, 2010.

Anderson, Jon Lee. "The Dangerous Absurdity of the Secret 'Cuban Twitter.'" *New Yorker,* April 4, 2014.

Angwin, Julia, Ariana Tobin, and Madeleine Varner. "Facebook (Still) Letting Housing Advertisers Exclude Users by Race." ProPublica, November 21, 2017.

Angwin, Julia, and Terry Parris Jr. "Facebook Lets Advertisers Exclude Users by Race." ProPublica, October 28, 2016.

Aschoff, Nicole. "America's Tipping Point?" In *Socialist Register 2019: A World Turned Upside Down.* Edited by Leo Panitch and Greg Albo. London: Merlin Press, 2018, 305–25.

Aschoff, Nicole. "The Glory Days Are Over." *Jacobin* 24, Journey to the Dark Side (Winter 2017).

Aschoff, Nicole. *The New Prophets of Capital.* London: Verso, 2015.

———. "The Smartphone Society." *Jacobin* 17, Ours to Master (Spring 2015).

Atkinson, Robert. "Relax. Robots Won't Cause Us to Run Out of Jobs." RealClearPolicy, January 24, 2019.

Autor, David, et al. "Concentrating on the Fall of the Labor Share." *American Economic Review: Papers and Proceedings* 107:5 (2017): 180–85.

Babwin, Don, and Micahel Tarm. "Officer Convicted of Murder in Slaying of Laquan McDonald." Associated Press, October 5, 2018.

Barker, Alex, and Mehreen Khan. "EU Fines Google Record 4.3bn over Android." *Financial Times,* July 18, 2018.

Barnes, Tom. "Saudi Arabia Prosecutor Says People Who Post Satire on Social Media Can Be Jailed." *Independent,* September 5, 2018.

Bell, Emily. "Facebook Is Eating the World." *Columbia Journalism Review*, March 7, 2016.

Belkhir, Lotfi, and Ahmed Elmeligi. "Assessing ICT Global Emissions Footprint: Trends to 2040 & Recommendations." *Journal of Cleaner Production* 177 (March 2018): 448–63.

Benjamin, Walter. "The Work of Art in the Age of Mechanical Reproduction." In *Illuminations: Essays and Reflections*. New York: Random House, 1968.

Bennett, Laura. "Kim Kardashian Just Wants to Be Seen. This 445-Page Book of Selfies Might Be Her Masterpiece." *Slate*, May 6, 2015.

Bernasek, Anna, and D. T. Morgan. *All You Can Pay: How Companies Use Our Data to Empty Our Wallets*. New York: Nation Books, 2015.

Blythe, Mark. *Austerity: The History of a Dangerous Idea*. New York: Oxford University Press, 2013.

Boltanski, Luc, and Eva Chiapello. *The New Spirit of Capitalism*. London: Verso, 2007.

Bourdieu, Pierre. "A Reasoned Utopia and Economic Fatalism." *New Left Review* 1/227 (January–February 1998).

———. *Distinction: A Social Critique of the Judgement of Taste*. Cambridge, MA: Harvard University Press, 2002.

Bowles, Nellie. "Silicon Valley Nannies Are Phone Police for Kids." *New York Times*, October 26, 2018.

———. "Tech CEOs Are in Love with Their Principal Doomsayer." *New York Times*, November 9, 2018.

———. "Clooneys to Attend a VIP Fundraiser for Clinton—and Sanders Fans Are Outraged." *New York Times*, April 12, 2016.

Bradshaw, Tim. "Google Chief Touts Utopian Ambitions." *Financial Times*, May 15, 2013.

Branch, Adam, and Zachariah Mampilly. *Africa Uprising: Popular Protest and Political Change*. London: Zed Books, 2015.

Brands, H. W. *American Colossus: The Triumph of Capitalism, 1865–1900*. New York: Anchor Books, 2010.

Braverman, Harry. *Labor and Monopoly Capital: The Degradation of Work in the Twentieth Century*. New York: Monthly Review Press, 1998.

Brooks, Rodney. "The Seven Deadly Sins of AI Prediction." *MIT Technology Review*, October 6, 2017.

Bruder, Jessica. *Nomadland: Surviving America in the Twenty-First Century*. New York: W. W. Norton, 2017.

Brynjolfsson, Eric, and Andrew McAfee. *The Second Machine Age: Work, Progress, and Prosperity in a Time of Brilliant Technologies*. New York: W. W. Norton, 2014.

Burgess, Jean, Alice Marwick, and Thomas Poell. *SAGE Handbook of Social Media*. Thousand Oaks, CA: Sage Publications, 2017.

Burns, Janet. "House Passes Bill That Could Have Teens Facing 15 Years for Trying to Sext." *Forbes*, June 2, 2017.

Cacioppo, John T., et al. "Marital Satisfaction and Break-ups Differ across On-line and Off-line Meeting Venues." *Proceedings of the National Academy of Sciences of the United States of America* 110, no. 25 (June 18): 10135–40.

Cagle, Matt. "Facebook, Instagram, and Twitter Provided Data Access for a Surveillance Product Marketed to Target Activists of Color." ACLU Northern California, October 11, 2016.

Cagle, Matt, and Nicole Ozer. "Amazon Teams Up with Government to Deploy Dangerous New Facial Recognition Technology." American Civil Liberties Union, May 22, 2018.

Cahill, Damien, and Martijn Konings. *Neoliberalism*. Medford, MA: Polity Press, 2017.

Caplan, Robyn, et al. "Algorithmic Accountability: A Primer." Report originally prepared for the Congressional Progressive Caucus's Tech Algorithm Briefing: How Algorithms Perpetuate Racial Bias and Inequality. *Data & Society*, April 18, 2018.

Caron, André H., and Letizia Caronia. *Moving Cultures: Mobile Communication in Everyday Life*. Montreal: McGill-Queen's University Press, 2007.

Carr, Nicholas. "How Smartphones Hijack Our Minds." *Wall Street Journal*, October 6, 2017.

———. *The Shallows: What the Internet Is Doing to Our Brains*. New York: W. W. Norton, 2011.

Chappell, Bill. "Philando Castile Is Remembered by St. Paul Public Schools: 'Kids Loved Him.'" *The Two-Way* (blog), NPR, July 7, 2016.

Charles, Ashley "Dotty." "As a Black, Gay Woman I Have to Be Selective in My Outrage. So Should You." *Guardian*, January 25, 2018.

Cherlin, Andrew J. *Labor's Love Lost: The Rise and Fall of the Working-Class Family in America*. New York: Russell Sage, 2014.

———. *The Marriage-Go-Round: The State of Marriage and the Family in America Today*. New York: Vintage Books, 2009.

Chopik, William J. "The Benefits of Social Technology Use Among Older Adults Are Mediated by Reduced Loneliness." *Cyberpsychology, Behavior, and Social Networking* 19, no. 9 (September 2016).

Cloos, Kassondra, and Julie Turkewitz. "Hundreds of Nude Photos Jolt Colorado School." *New York Times*, November 6, 2015.

Collins, Ben, and Joseph Cox. "Jenna Abrams, Russia's Clown Troll Princess, Duped the Mainstream Media and the World." *Daily Beast*, November 2, 2017.

Colvin, Alexander J. S. "The Growing Use of Mandatory Arbitration." Report. Economic Policy Institute, April 6, 2018, www.epi.org/publication/the-growing-use-of-mandatory-arbitration-access-to-the-courts-is-now-barred-for-more-than-60-million-american-workers/.

Corbin, Kenneth. "HUD Is Suing Facebook for Housing Discrimination." *Forbes*, March 29, 2019.

Cubarrubia, R. J. "Beyonce Calls Herself a 'Modern-Day Feminist.'" *Rolling Stone*, April 3, 2013.

Curran, Dylan. "Are You Ready? Here Is All the Data Facebook and Google Have on You." *Guardian*, March 30, 2018.

Deleuze, Gilles. "Postscript on the Societies of Control." *L'Autre journal*, no. 1 (May 1990), https://theanarchistlibrary.org/library/gilles-deleuze-postscript-on-the-societies-of-control.

Desai, Bhairavi. "How to Beat Uber." Interview by Christ Brooks. *Jacobin*, August 19, 2018.

Dewan, Shaila, and Richard A. Oppel Jr. "In Tamir Rice Case, Many Errors by Cleveland Police, Then a Fatal One." *New York Times*, January 22, 2015.

Diakopoulos, Nicholas, and Sorelle Friedler. "How to Hold Algorithms Accountable." *MIT Technology Review*, November 17, 2016.

Dixon, Pam, and Robert Gellman. *The Scoring of America: How Secret Consumer Scores Threaten Your Privacy and Your Future*. Report. World Privacy Forum, April 2, 2014.

Dobson, Jim. "Inside the Verdura Resort in Sicily, Home to Google's Top Secret Summer Camp." *Forbes*, July 28, 2017.

Fang, Lee. "The CIA Is Investing in Firms That Mine Your Tweets and Instagram Photos." *The Intercept*, April 14, 2016.

Federici, Silvia. *Caliban and the Witch: Women, the Body and Primitive Accumulation*. Brooklyn, NY: Autonomedia, 2014.

Feng, Emily. "Beijing Widens Control of Wildly Popular Short-Video Apps." *Financial Times*, January 12, 2019.

Ferguson, Thomas, Paul Jorgensen, and Jie Chen. "Industrial Structure and Party Competition in the Age of Hunger Games: Donald Trump and the 2016 Presidential Election." Working Paper No. 66. Institute for New Economic Thinking, January 2018.

Fitzsimmons, Emma G. "Uber Hit with Cap as New York City Takes Lead in Crackdown." *New York Times*, August 8, 2018.

Foroohar, Rana. "Release Big Tech's Grip on Power." *Financial Times*, June 18, 2017.

Fraser, Nancy. "From Progressive Neoliberalism to Trump and Beyond." *American Affairs* 1, no. 4 (Winter 2017).

Freed, Richard. *Wired Child: Reclaiming Childhood in a Digital Age*. Self-published, CreateSpace Independent Publishing Platform, 2015.

Freedom on the Net 2016. "Silencing the Messenger: Communication Apps Under Pressure." Freedom House, November 2016.

Fried, Ina. "What Facebook Knows About You." Axios, January 2, 2019.

Friedman, Richard A. "The Big Myth About Teenage Anxiety." *New York Times*, September 7, 2018.

Friedman, Thomas L. *The World Is Flat: A Brief History of the Twenty-First Century*. New York: Farrar, Straus and Giroux, 2005.

Gabler, Neal. *Life: The Movie—How Entertainment Conquered Reality*. New York: Vintage Books, 2000.

Galloway, Scott. *The Four: The Hidden DNA of Amazon, Apple, Facebook, and Google*. New York: Penguin Books, 2017, Kindle edition.

Gereffi, Gary. "Beyond the Producer-Driven/Buyer-Driven Dichotomy: The Evolution of Global Value Chains in the Internet Era." *IDS Bulletin* 32, no. 3 (2001).

Gertner, Jon. *The Idea Factory: Bell Labs and the Great Age of American Innovation*. New York: Penguin Books, 2012.

Ghose, Anindya. *Tap: Unlocking the Mobile Economy*. Cambridge, MA: MIT Press, 2017.

Gillis, Tom. "Wikileaks Only Told You Half the Story—Why Encryption Matters More Than Ever." *Forbes*, March 21, 2017.

Goffman, Erving. *The Presentation of Self in Everyday Life*. New York: Anchor Books, 1959.

Goodman, J. David. "Amazon Pulls Out of Planned New York City Headquarters." *New York Times*, February 14, 2019. Gosselin, Peter. "IBM Accused of Violating Federal Anti-Age Discrimination Law." ProPublica, March 27, 2019.

Gottfried, Jeffrey, and Elisa Shearer. "News Use Across Social Media Platforms 2017." Pew Research Center, September 7, 2017.

Graeber, David. "Of Flying Cars and the Declining Rate of Profit." *Baffler* 19, March 2012.

Graham, Stephen. *Cities Under Siege: The New Military Urbanism*. London: Verso, 2011.

Grandin, Greg. *Fordlandia: The Rise and Fall of Henry Ford's Forgotten Jungle City*. New York: Metropolitan Books, 2009.

Greenfield, Adam. *Radical Technologies: The Design of Everyday Life*. London: Verso, 2017.

Gregory, Alice. "This Startup Wants to Neutralize Your Phone—And Un-Change the World." *Wired*, January 16, 2018.

Habermas, Jürgen. *The Structural Transformation of the Public Sphere: An Inquiry into a Category of Bourgeois Society*. Cambridge, MA: MIT Press, 1989. Cited in McChesney, *Digital Disconnect*.

Harari, Yuval. *21 Lessons for the 21st Century*. New York: Spiegel & Grau, 2018, Kindle edition.

Haraway, Donna J. "A Cyborg Manifesto: Science, Technology, and Socialist Feminism in the Late Twentieth Century." *Simians, Cyborgs, and Women: The Reinvention of Nature*. London: Free Association Books, 1996.

Harcourt, Bernard E. *Exposed: Desire and Disobedience in the Digital Age*. Cambridge, MA: Harvard University Press, 2015.

Harris, Mark. "Inside the First Church of Artificial Intelligence." *Wired*, November 15, 2017.

Hasinoff, Amy Adele. *Sexting Panic: Rethinking Criminalization, Privacy, and Consent*. Urbana: University of Illinois Press, 2015.

Hassan, Adeel. "Family of Jordan Edwards Says 15 Years Is Not Enough for Officer Who Murdered Him." *New York Times*, August 31, 2018.

Hawkins, Andrew J. "Uber Sues to Overturn New York City's Cap on New Ride-Hail Drivers." *The Verge*, February 15, 2019.

Helft, Miguel. "Meet the Man Who Derailed Facebook's Plan to Provide Free Internet in India." *Forbes*, February 25, 2016.

Hillis, Ken. *Online a Lot of the Time: Ritual, Fetish, Sign*. Durham, NC: Duke University Press, 2009.

Hochschild, Arlie Russell. *The Time Bind: When Work Becomes Home and Home Becomes Work*. New York: Metropolitan Books, 1997.

Hodgson, Camila. "Inside the Secretive World of Stalking Apps." *Financial Times*, July 18, 2019.

Hollister, Sean. "Facebook Is Officially Building an Internet Satellite: Athena." C/Net, July 20, 2018.

Hook, Leslie. "The Software Engineer Who Lifted the Lid on Sexual Harassment at Uber and Inspired Women to Speak Out." *Financial Times*, December 11, 2017.

Hook, Leslie, and Hannah Kuchler. "Uber's Turmoil Compounded by David Bonderman's Sexist Quip." *Financial Times*, June 14, 2017.

Hu, Tung-Hui. *A Prehistory of the Cloud*. Cambridge, MA: MIT Press, 2016.

Illouz, Eva. *Cold Intimacies: The Making of Emotional Capitalism*. Cambridge, MA: Polity Press, 2007.

Jaffe, Sarah. *Necessary Trouble: Americans in Revolt*. New York: Nation Books, 2016.

James, Selma. *Sex, Race, and Class: The Perspective of Winning—A Selection of Writings, 1952–2011*. Oakland, CA: PM Press, 2012.

Jameson, Frederic. *Archaeologies of the Future: The Desire Called Utopia and Other Science Fictions*. London: Verso, 2007.

———. "The End of Temporality." *Critical Inquiry* 29 (2003): 4.

———. *Postmodernism, or, the Cultural Logic of Late Capitalism*. Durham, NC: Duke University Press, 1991.

Jasper, Simone. "He Took Video Outside an ICE Raid in Sanford. Now He's Charged with Communicating Threats." Raleigh (NC) *News & Observer*, February 7, 2019.

Kantor, Jodi, and Megan Twohey. "Harvey Weinstein Paid Off Sexual Harassment Accusers for Decades." *New York Times*, October 5, 2017.

Kantor, Jodi, and David Streitfeld. "Inside Amazon: Wrestling Big Ideas in a Bruising Workplace." *New York Times*, August 15, 2015.

Kardashian, Kim. *Selfish*. New York: Rizzoli, 2016.

Kavanagh, Jennifer, and Michael D. Rich. "Truth Decay: An Initial Exploration of the Diminishing Role of Facts and Analysis in American Public Life." Santa Monica: Rand Corporation, 2018.

Kazmin, Amy. "'I Am a Troll' by Swati Chaturvedi." *Financial Times*, February 20, 2017.

Kellaway, Lucy. "The Bliss of Being 396 Miles from My Lost Smartphone." *Financial Times*, September 25, 2016.

Khan, Lina M. "Amazon's Antitrust Paradox." *Yale Law Journal* 126, no. 3 (January 2017).

King, Jamilah. "#blacklivesmatter: How Three Friends Turned a Spontaneous Facebook Post into a Global Phenomenon." *California Sunday Magazine*, 2015.

Krugman, Paul. "Why Most Economists' Predictions Are Wrong," *Red Herring*, June 1998.

Kuchler, Hannah. "US Midterms: Democrats Look to Big Data to Beat Trump." *Financial Times*, November 1, 2018.

Lanier, Jaron. *Ten Arguments for Deleting Your Social Media Accounts Right Now*. New York: Henry Holt, 2018.

Lareau, Annette. *Unequal Childhoods: Class, Race, and Family Life*. Berkeley: University of California Press, 2011.

Laufer, William S. "Social Accountability and Corporate Greenwashing." *Journal of Business Ethics* 43, no. 3 (2003): 43.

Laughland, Oliver, and Amanda Holpuch. "Walter Scott: Large Crowd Attends Funeral of Man Shot by Police Officer." *Guardian*, April 11, 2015.

Lears, Jackson. *Rebirth of a Nation: The Making of Modern America, 1877–1920*. New York: Harper Perennial, 2010.

Lecher, Colin. "An Internal Google Email Shows How the Company Cracks Down on Leaks." *Verge*, May 22, 2017.

Lee, Joonkoo, and Gary Gereffi. "The Co-Evolution of Concentration in Mobile Phone Global Value Chains and Its Impact on Social Upgrading in Developing Countries." Capturing the Gains Working Paper 25. March 23, 2013. Available for download at SSRN, https://papers.ssrn.com/sol3/papers.cfm?abstract_id=2237510.

Lee, Min Kyung, et al. "Working with Machines: The Impact of Algorithmic and Data-Driven Management on Human Workers." Carnegie Mellon University, 2015.

Lee, Timothy B. "Google Is Losing Allies Across the Political Spectrum." *Ars Technica*, August 31, 2017.

Leetaru, Kalev. "Geofeedia Is Just the Tip of the Iceberg: The Era of Social Surveillance." *Forbes*, October 12, 2016.

Leng, Yan, Dominiquo Santistevan, and Alex "Sandy" Pentland. "Familiar Strangers: The Collective Regularity in Human Behaviors." arXiv.org e-print archive, June 12, 2018, https://arxiv.org/pdf/1803.08955.pdf.

Leswing, Kif. "The Average iPhone Is Unlocked 80 Times Per Day." *Business Insider*, April 18, 2016.

Levin, Sam. "Officer Punched Oscar Grant and Lied About Facts in 2009 Killing, Records Show." *Guardian*, May 2, 2019.

———. "Police Shot a Pregnant California Teen—but with No Video, the Case Dried Up." *Guardian*, March 15, 2018.

———. "Google Accused of 'Extreme' Gender Pay Discrimination by US Labor Department." *Guardian*, April 7, 2017.

———. "Sexual Harassment and the Sharing Economy: The Dark Side of Working for Strangers." *Guardian*, August 23, 2017.

Levine, Yasha. *Surveillance Valley: The Secret Military History of the Internet*. New York: Hachette Books, 2017.

Lewis, Leo, and Emma Jacobs. "How Business Is Capitalising on the Millennial Instagram Obsession." *Financial Times*, July 13, 2018.

Lewis, Paul. "'Our Minds Can Be Hijacked': The Tech Insiders Who Fear a Smartphone Dystopia." *Guardian*, October 6, 2017.

Lewis, Taylor. "History Makers: Millions March NYC Organizers Synead Nichols and Umaara Elliot." *Essence*, February 18, 2015.

Licoppe, Christian. "'Connected' Presence: The Emergence of a New Repertoire for Managing Social Relationships in a Changing Communication Technoscape." *Environment and Planning D: Society and Space* 22, no. 1 (2004): 150.

Ling, Rich. *New Tech, New Ties: How Mobile Communication Is Reshaping Social Cohesion*. Cambridge, MA: MIT Press, 2008.

Ling, Rich, and Jonathan Donner. *Mobile Communication*. Hoboken: John Wiley, 2013.

Losse, Katherine. *The Boy Kings: A Journey into the Heart of the Social Network*. New York: Free Press, 2012.

Lucero, Leanne. "Safe Spaces in Online Places: Social Media and LGBTQ Youth." *Multicultural Education Review* 9, no. 2 (April 2017): 117–28.

Luckerson, Victor. "The Rise of the Like Economy." *Ringer*, February 15, 2017.

Lum, Cynthia, Megan Stoltz, Christopher S. Koper, and J. Amber Scherer. "Research on Body-Worn Cameras: What We Know, What We Need to Know." *Criminology & Public Policy* (2019): 18.

Lynch, Kevin. *The Image of the City*. Cambridge, MA: Publication of the Joint Center for Urban Studies, 1960.

Lynd, Robert S., and Helen Merrell Lynd. *Middletown: A Study in Contemporary American Culture*. London: Constable, 1929.

Macauley, Thomas. "What Google Knows About You—And How to Make It Forget." *TechWorld*, November 1, 2018.

MacKenzie, Donald, and Judy Wajcman, eds. *The Social Shaping of Technology*. 2nd ed. Reprint, Buckingham, UK: McGraw-Hill Education/Open University Press, 2006.

Margolis, Jonathan. "A Crash Course in Technology for Ageing and Infirm Bodies." *Financial Times*, November 9, 2016.

Marriage, Madison, and Richard Waters. "Alphabet Set to Face Down Shareholder Dissent." *Financial Times*, June 4, 2017.

Mayer-Schönberger, Victor, and Kenneth Cukier. *Big Data: A Revolution That Will Transform How We Live, Work, and Think*. London: John Murray, 2013.

Mazzucato, Mariana. *The Entrepreneurial State: Debunking Public vs. Private Sector Myths*. New York: Public Affairs, 2015.

McCarthy, Justin. "Americans Maintain a Positive View of Bernie Sanders." Gallup, "Politics," October 5, 2018.

McCarthy, Niall. "China Now Boasts More Than 800 Million Internet Users and 98% of Them Are Mobile." *Forbes*, August 23, 2018.

McChesney, Robert W. *Digital Disconnect: How Capitalism Is Turning the Internet Against Democracy*. New York: New Press, 2013.

McCracken, Scott. "Cyborg Fictions: The Cultural Logic of Posthumanism." In *Ruthless Criticism of All That Exists, Socialist Register 1997*, vol. 33. Edited by Leo Panitch. London: Merlin Press, 1996.

McKinsey Global Institute. "China's Digital Economy: A Leading Global Force." McKinsey & Company, Discussion Paper, August 2017.

Melendez, Steven, and Alex Pasternack. "Here Are the Data Brokers Quietly Buying and Selling Your Personal Information." *Fast Company*, March 2, 2019.

Merrill, Jeremy B., and Ariana Tobin. "Facebook Moves to Block Ad Transparency Tools—Including Ours." ProPublica, January 28, 2019.

Meyerson, Collier. "The Founders of Black Lives Matter: 'We Gave Tongue to Something That We All Knew Was Happening.'" *Glamour*, Glamour Women of the Year, November 1, 2016.

Milgram, Stanley, and Thomas Blass. *The Individual in a Social World: Essays and Experiments*. Orig. 1977. London: Pinter & Martin, 2010.

Mitchell, Stacy. "Amazon Doesn't Just Want to Dominate the Market—It Wants to Become the Market." *Nation*, February 15, 2018.

Moazed, Alex, and Nicholas L. Johnson. *Modern Monopolies: What It Takes to Dominate the 21st-Century Economy*. New York: St. Martin's Press, 2016.

Molla, Rani. "Google Leads the World in Digital and Mobile Ad Revenue." *Vox*, July 24, 2017.

Moody, Kim. *On New Terrain: How Capital Is Reshaping the Battleground of Class War*. Chicago: Haymarket Books, 2017.

Moore, Jason W. *Capitalism in the Web of Life: Ecology and the Accumulation of Capital*. London: Verso, 2015.

Morozov, Evgeny. "From Slacktivism to Activism." *Foreign Policy*, September 5, 2009.

Mozur, Paul. "A Genocide Incited on Facebook, with Posts from Myanmar's Military." *New York Times*, October 15, 2018.

Mudde, Cas. "Don't Be Fooled. The Midterms Were Not a Bad Night for Trump." *Guardian*, November 7, 2018.

Mullins, Brody, and Jack Nicas. "Paying Professors: Inside Google's Academic Influence Campaign." *Wall Street Journal*, July 14, 2017.

Mullins, Brody, Rolfe Winkler, and Brent Kendall. "Inside the U.S. Antitrust Probe of Google." *Wall Street Journal*, March 19, 2015.

Munro, Alice. *A Wilderness Station: Selected Stories, 1968–1994*. New York: Vintage International, 1997.

Murgia, Madhumita, and Oliver Ralph. "Facebook Blocks Admiral from Using Posts to Assess Drivers." *Financial Times*, November 2, 2016.

Nakashima, Ryan. "AP Exclusive: Google Tracks Your Movements, Like It or Not." *AP News*, August 13, 2018.

Newcomer, Eric. "In Video Uber CEO Argues with Driver over Falling Fares." *Bloomberg*, February 28, 2017.

Nicas, Jack. "How YouTube Drives People to the Internet's Darkest Corners." *Wall Street Journal*, February 7, 2018.

Noble, David. *Forces of Production: A Social History of Industrial Automation*. New Brunswick, NJ: Transaction, 2011.

O'Connor, Sarah. "When Your Boss Is an Algorithm." *Financial Times*, September 8, 2016.

O'Gieblyn, Meghan. "Ghost in the Cloud: Transhumanism's Simulation Theology." *n + 1* (*Half-Life*) 28 (Spring 2017).

O'Neil, Cathy. *Weapons of Math Destruction: How Big Data Increases Inequality and Threatens Democracy*. New York: Crown, 2016.

O'Neil, Luke. "My Candid Conversations with Extremely Online Folks Who Suffer from Internet Broken Brain." *Esquire*, March 21, 2018.

O'Reilly, Holly Figueroa. "Trump Blocked Me on Twitter. Not Any More." *Guardian*, May 24, 2018.

Orenstein, Peggy. *Girls & Sex: Navigating the Complicated New Landscape*. New York: HarperCollins, 2016.

Oxford, Andrew. "California Lawmakers Block Expansion of Data Privacy Law." *AP News*, May 16, 2019.

Panitch, Leo, and Sam Gindin. *The Making of Global Capitalism: The Political Economy of American Empire*. London: Verso, 2012.

Pariser, Eli. *The Filter Bubble: How the New Personalized Web Is Changing What We Read and How We Think*. New York: Penguin, 2011.

Piketty, Thomas, Emmanuel Saez, and Gabriel Zucman. "Distributional National Accounts: Methods and Estimates for the United States." NBER Working Paper 22945. Cambridge, MA: National Bureau of Economic Research, December 2016.

Pine, Joseph B., and James H. Gilmore. "Welcome to the Experience Economy." *Harvard Business Review*, July–August 1998.

Polletta, Francesca. *It Was Like a Fever: Storytelling in Protest and Politics*. Chicago: University of Chicago Press, 2006.

Pollin, Robert. "De-Growth vs. A Green New Deal." *New Left Review* 112 (July–August 2018).

Popper, Nathaniel. "A Cryptocurrency Millionaire Wants to Build a Utopia in Nevada." *New York Times*, November 1, 2018.

Posner, Eric A., and E. Glen Weyl. "Want Our Personal Data? Pay for It." *Wall Street Journal*, April 20, 2018.

Postman, Neil. *The Disappearance of Childhood*. New York: Vintage, 2011, Kindle edition.

Price, Catherine. "How to Break Up with Your Phone." *New York Times*, February 13, 2018.

Przybylski, Andrew K., and Netta Weinstein. "A Large Scale Test of the Goldilocks Hypothesis: Quantifying the Relations Between Digital Screens and the Mental Well-Being of Adolescents." *Psychological Science* 28, no. 2 (January 2017).

Purnell, Newley. "The Internet Is Filling Up Because Indians Are Sending Millions of 'Good Morning!' Texts." *Wall Street Journal*, January 22, 2018.

Putnam, Robert D. *Bowling Alone: The Collapse and Revival of American Community*. New York: Simon & Schuster, 2000.

Ralph, Oliver. "Insurance and the Big Data Technology Revolution." *Financial Times*, February 24, 2017.

Ransby, Barbara. *Making All Black Lives Matter: Reimagining Freedom in the 21st Century*. Berkeley: University of California Press, 2018.

Rivoli, Dan. "New York City Will Propose Minimum Wage for Uber, App Drivers." *New York Daily News*, July 2, 2018.

Robb, Michael B. "The New Normal: Parents, Teens, Screens, and Sleep in the United States." San Francisco, CA: Common Sense Media, 2019.

Rojek, Chris. *Presumed Intimacy: Parasocial Interaction in Media, Society and Celebrity Culture*. Cambridge, MA: Polity Press, 2016.

———. *Celebrity*. London: Reaktion Books, 2001.

Romero, Luiz. "US Terror Attacks Are Increasingly Motivated by Right-Wing Views." *Quartz*, October 24, 2018.

Roose, Kevin. "After Uproar, Instacart Backs Off Controversial Tipping Policy." *New York Times*, February 6, 2019.

———. "Forget Washington. Facebook's Problems Abroad Are Far More Disturbing." *New York Times*, October 29, 2017.

Rosin, Hanna. "Why Kids Sext." *Atlantic*, November 2014.

Roy, Nilanjana. "Why the Constant Struggle to Manage Our Inboxes Is About More Than Just Work." *Financial Times*, May 2, 2017.

Rudder, Christian. *Dataclysm: Who We Are* When We Think No One's Looking*. New York: Crown, 2014.

Rushe, Dominic. "Scholar Says Google Criticism Cost Him Job: 'People Are Waking Up to Its Power.'" *Guardian*, August 31, 2017.

Safdar, Khadeeja. "On Hold for 45 Minutes? It Might Be Your Secret Customer Score." *Wall Street Journal*, November 1, 2018.

Sainato, Michael. "'We Are Not Robots': Amazon Warehouse Employees Push to Unionize." *Guardian*, January 1, 2019.

Sales, Nancy Jo. *American Girls: Social Media and the Secret Lives of Teenagers*. New York: Knopf, 2016.

———. "Tinder and the Dawn of the 'Dating Apocalypse.'" *Vanity Fair*, August 6, 2015.

Sandberg, Sheryl. *Lean In: Women, Work, and the Will to Lead*. New York: Knopf, 2013.

Sauri, Emilio. "Cognitive Mapping, Then and Now: Postmodernism, *Indecision*, and American Literary Globalism." *Twentieth-Century Literature* 57, no. 3–4 (Fall–Winter 2011): 472–91.

Scahill, Jeremy. *The Assassination Complex: Inside the Government's Secret Drone Warfare Program*. New York: Simon & Schuster, 2016.

Schmidt, Charles. "New Studies Link Cell Phone Radiation with Cancer." *Scientific American*, March 29, 2018.

Schmidt, Janek. "Always Practise Safe Text: The German Traffic Light for Smartphone Zombies." *Guardian*, April, 29, 2016.

Schneier, Bruce. *Data and Goliath: The Hidden Battles to Collect Your Data and Control Your World*. New York: W. W. Norton, 2016.

Scholz, Trebor. "Platform Cooperativism vs. the Sharing Economy." *Medium*, December 5, 2014.

Schüll, Natasha Dow. "Stuck in the Machine Zone: Your Sweet Tooth for 'Candy Crush.'" *All Tech Considered*, NPR, June 7, 2014.

———. *Addiction by Design: Machine Gambling in Las Vegas*. Princeton, NJ: Princeton University Press, 2012.

Semuels, Alana. "How Amazon Helped Kill a Seattle Tax on Business." *Atlantic*, June 13, 2018.

Siegel, Rachel. "DoorDash to Change Its Controversial Tipping Policy After Outcry." *Washington Post*, July 24, 2019.

Silver, Beverly J. "Workers of the World." *Jacobin* (*Rank and File*) 22 (Summer 2016).

———. *Forces of Labor: Workers' Movements and Globalization Since 1870*. Cambridge, UK: Cambridge University Press, 2003.

Simmons, Lee. "How Apple—and the Rest of Silicon Valley—Avoids the Tax Man." *Wired*, August 30, 2016.

Singer, Natasha. "How Google Took Over the Classroom." *New York Times*, May 13, 2017.

Sledge, Matt. "CIA's Gus Hunt on Big Data: We 'Try to Collect Everything and Hang On to It Forever.'" *Huffington Post*, March 20, 2013.

Snowden, Edward. "Foreword: Elected by Circumstance." In *The Assassination Complex: Inside the Government's Secret Drone Warfare Program*. Edited by Jeremy Scahill. New York: Simon & Schuster, 2016.

Solnit, Rebecca. "Standing Rock Inspired Ocasio-Cortez to Run. That's the Power of Protest." *Guardian*, January 14, 2019.

Solon, Olivia, and Sabrina Siddiqui. "Forget Wall Street—Silicon Valley Is the New Political Power in Washington." *Guardian*, September 3, 2017.

Soper, Taylor. "Smartphone Shopping: Amazon Says Thanksgiving Mobile Orders Spiked 50% from Last Year." GeekWire, November 24, 2017.

Spigel, Lynn. *Make Room for TV: Television and the Family Ideal in Postwar America*. Chicago: University of Chicago Press, 1992.

Srivastava, Mehul. "Seven-Year-Old Bana Al-Abed, the 'Face of Aleppo.'" *Financial Times*, March 10, 2017.

Srnicek, Nick. *Platform Capitalism*. Cambridge, MA: Polity Press, 2017.

Stallman, Richard. "Talking to the Mailman." Interview by Rob Lucas. *New Left Review* 113 (September–October 2018).

———. "A Radical Proposal to Keep Your Personal Data Safe." *Guardian*, April 3, 2018.

Stampler, Laura. "Amazon Will Pay a Whopping $0 in Federal Taxes on $11.2 Billion Profits." *Fortune*, February 14, 2019.

Statt, Nick. "Google Confirms the End of Its Module Project Ara Smartphone." *The Verge*, September 2, 2016.

Strain, Garrett, Eli Moore, and Sami Gambhir. "AT&T's Digital Divide in California: An Analysis of AT&T Fiber Deployment and Wireline Broadband Speeds in California." Policy brief. Berkeley: University of California, Haas Institute for a Fair and Inclusive Society, 2017.

Sullivan, Andrew. "I Used to Be a Human Being." Intelligencer, *New York*, September 18, 2016.

Sundararajan, Arun. *The Sharing Economy: The End of Employment and the Rise of Crowd-Based Capitalism*. Cambridge, MA: MIT Press, 2016.

Sunstein, Cass R. *#Republic: Divided Democracy in the Age of Social Media*. Princeton, NJ: Princeton University Press, 2017, Kindle edition.

Taplin, Jonathan. *Move Fast and Break Things: How Facebook, Google, and Amazon Cornered Culture and Undermined Democracy*. Boston: Little, Brown, 2017.

———. "Google's Disturbing Influence over Think Tanks." *New York Times*, August 30, 2017.

Taylor, Astra. *The People's Platform: Taking Back Power and Culture in the Digital Age*. New York: Picador, 2014.

Taylor, Keeanga-Yamahtta. *From #BlackLivesMatter to Black Liberation*. Chicago: Haymarket Books, 2014.

Temin, Peter. *The Vanishing Middle Class: Prejudice and Power in a Dual Economy*. Cambridge, MA: MIT Press, 2017.

Thiel, Peter. *Zero to One: Notes on Startups, or How to Build the Future*. New York: Crown, 2014.

Thomas, Daniel. "Digitally Weary Users Switch to 'Dumb' Phones." *Financial Times*, February 22, 2016.

Thomas, Minu, and Sun Sun Lim. "On Maids, Mobile Phones, and Social Capital— ICT Use by Female Migrant Workers in Singapore and Its Policy Implication." In *Mobile Communication and Social Policy*. Edited by J. Katz. New Brunswick, NJ: Transaction, 2010.

Thompson, Clive. *Smarter Than You Think: How Technology Is Changing Our Minds for the Better*. New York: Penguin, 2013.

Tiku, Nitasha. "Google Deliberately Confuses Its Employees, Fed Says." *Wired*, July 25, 2017.

Tobin, Ariana. "HUD Sues Facebook over Housing Discrimination and Says the Company's Algorithms Have Made the Problem Worse." ProPublica, March 28, 2019.

Toplensky, Rochelle. "EU Fines Google €2.4bn over Abuse of Search Dominance." *Financial Times*, June 27, 2017.

Torre, Giovanni. "Indigenous Australians Use Tech to Expose Police Abuse." *New York Times*, August 14, 2018.

Trachtenberg, Alan. *The Incorporation of America: Culture and Society in the Gilded Age*. New York: Hill & Wang, 2007.

Tufekci, Zeynep. "YouTube, the Great Radicalizer." *New York Times*, March 10, 2018.

———. *Twitter and Tear Gas: The Power and Fragility of Networked Protest*. New Haven, CT: Yale University Press, 2017.

Turkle, Sherry. *Reclaiming Conversation: The Power of Talk in a Digital Age*. New York: Penguin, 2015.

Twenge, Jean M. "Have Smartphones Destroyed a Generation?" *Atlantic*, September 2017.

Ullman, Ellen. *Life in Code: A Personal History of Technology*. New York: Farrar, Straus & Giroux, 2017.

———. *Close to the Machine: Technophilia and Its Discontents*. New York: Picador, 1997.

Vigneri, Luigi, et al. "Taming the Android AppStore: Lightweight Characterization of Android Applications." Eurecom Research Report RR-15-305, April, 27 2015.

Viljoen, Salome. "Facebook's Surveillance is Nothing Compared with Comcast, AT&T and Verizon." *Guardian*, April 6, 2018.

Vincent, James. "Former Facebook Exec Says Social Media Is Ripping Apart Society." *Verge*, December 11, 2017.

Vogelstein, Fred. *Dogfight: How Apple and Google Went to War and Started a Revolution*. New York: Sara Crichton Books, 2013.

Waddell, Kaveh. "Why Bosses Can Track Their Employees 24/7." *Atlantic*, January 6, 2017.

Wagner, Kurt. "Facebook to Raise Pay for Thousands of Contract Workers, Including Content Moderators." *Bloomberg*, May 13, 2019.

Wajcman, Judy. *Pressed for Time: The Acceleration of Life in Digital Capitalism*. Chicago: University of Chicago Press, 2016.

Wang, Maya. "China's Chilling 'Social Credit' Blacklist." *Wall Street Journal*, December 11, 2017.

Waters, Richard, Rochelle Toplensky, and Aliya Ram. "Brussels' €2.4bn Fine Could Lead to Damages Cases and Probes in Other Areas of Search." *Financial Times*, June 28, 2017.

Weil, David. *The Fissured Workplace: Why Work Became So Bad for So Many and What Can Be Done to Improve It*. Cambridge, MA: Harvard University Press, 2014.

Weiss, Richard. *The American Myth of Success: From Horatio Alger to Norman Vincent Peale*. New York: Basic Books, 1969.

Whittaker, Meredith, et al. *AI Now Report 2018*. New York: New York University, AI Now Institute, December 2018.

Willingham, Daniel T. "Smartphones Don't Make Us Dumb." *New York Times*, January 20, 2015.

Wilson, Jason. "Who Are the Proud Boys, 'Western Chauvinists' Involved in Political Violence?" *Guardian*, July 14, 2018.

Wilson, Julie, and Emily Chivers Yochim. "Pinning Happiness: Affect, Social Media, and the Work of Mothers." In *Cupcakes, Pinterest, and Lady Porn: Feminized Popular Culture in the Early Twenty-First Century*. Edited by Elana Levine. Urbana: University of Illinois Press, 2015.

Winkie, Luke. "Why Men Are Paying Women for Accurate Critiques of Their Dicks." *MEL Magazine*, 2018.

Winner, Langdon. "Do Artifacts Have Politics?" *Daedalus* (Modern Technology: Problem or Opportunity?) 109, no. 1 (Winter 1980).

Witt, Emily. "How the Survivors of Parkland Began the Never Again Movement." *New Yorker*, February 19, 2018.

———. "The Optimistic Activists for a Green New Deal: Inside the Youth-Led Singing Sunrise Movement." *New Yorker*, December 23, 2018.

Woetzel, Jonathan, et al. "China's Digital Economy: A Leading Global Force." Report. McKinsey & Company, McKinsey Global Institute, August 2017.

Woyke, Elizabeth. *The Smartphone: Anatomy of an Industry*. New York: New Press, 2014.

Wu, Tim. *The Attention Merchants: The Epic Scramble to Get Inside Our Heads*. New York: Knopf, 2016, Kindle edition.

Yang, Yuan. "Beijing Now Able to Flag Weibo Posts as 'Rumour.'" *Financial Times*, November 5, 2018.

———. "Apple Investigates Illegal Student Labour at Watch Assembly Plant." *Financial Times*, October 28, 2018.

Zuboff, Shoshana. *The Age of Surveillance Capitalism: The Fight for a Human Future at the New Frontier of Power*. New York: PublicAffairs, 2019.

Zucman, Gabriel. "Global Wealth Inequality." NBER Working Paper 25462. Cambridge, MA: National Bureau of Economic Research, January 2019.

———. "How Corporations and the Wealthy Avoid Taxes (and How to Stop Them)." *New York Times*, November 10, 2017.

Index

"Abrams, Jenna," 109
ACLU (American Civil Liberties Union), 21, 22, 96, 148–49, 153, 176n28
Acxiom, 72, 77
addiction to social media, 65–69
Admiral, 78, 79
Adolfsson, Martin, 84
AdSense, 52
Advanced Research Projects Agency Network (ARPANET), 80
advertising: data collection and, 76–77, 83–84; on Facebook, 47–48; on Google, 52; political, 149
AdWords, 52
Afghanistan, politics in, 95–96
age discrimination, 149
AI Now, 129, 157
Airbnb, 42, 119, 120, 148
al-Abed, Bana, 90
al-Abed, Fatemah, 90
Aldridge, Rasheen, 91
algorithm(s): for consumer categories, 77; in dystopian future, 129–30, 131; in politics, 109–10; recommender, 67; search, 51–52, 53
algorithmic accountability, 151, 157–59
"algorithmic management," 32–34
Alibaba, 4, 42
Allard, LaDonna Brave Bull, 104
aloneness, 7
Alphabet, 41, 55, 76, 122, 150
Alpha Go Zero program, 122
alter-globalization activists, 91
Amazon: acquisitions by, 41; and CIA, 81; in Europe, 150; marketplace model

of, 150; as monopoly, 53–54; and new capitalism, 118–19; as new titan, 38–41, 44, 45; power of, 54–55; Rekognition software of, 149; tax evasion by, 49; warehouse workers at, 31–32, 33, 34; working conditions at, 46
Amazon Shopping, 30
Amazon Web Services (AWS), 41
American Academy of Pediatrics, 9
American Association of People with Disabilities, 55
American Civil Liberties Union (ACLU), 21, 22, 96, 148–49, 153, 176n28
American Dream, 120–21
American Library Association, 55
amplification, 91
analog networks, in cognitive mapping, 143
Android operating system, 39, 41, 53, 69, 70–71, 72
"angel investor," 120
Ant Financial, 42
antidiscrimination laws, 149
antitrust laws and practices, 43–44, 52–53, 55, 150
AOL, data collection from, 81
app(s): top ten, 38
"app dashboard," 69
app jobs, 30–34, 35, 137
Apple: data collection from, 81; in Europe, 150; low-paid workers at, 147; as new titan, 42; refusal of "right to repair" by, 155; supply chain of, 28–29; tax evasion by, 49–50; working conditions at, 46